TEXT BOOK
OF
HEAT

DPH PHYSICS SERIES

TEXT BOOK OF HEAT

By

D.K. Jha

DISCOVERY PUBLISHING HOUSE
NEW DELHI-110002

Published by:
Namit Wasan

DISCOVERY PUBLISHING HOUSE PVT. LTD.
4383/4B, Ansari Road, Darya Ganj
New Delhi-110 002 (India)
Phone : +91-11-23279245; 23253475; 43596065
E-mail : discoverybooksindia@gmail.com
discoverypublishinghouse@gmail.com
namitwasan9@gmail.com
web : www.discoverypublishinggroup.com

***Edition:* 2020**

ISBN: 978-81-7141-911-1

Text Book of Heat

Printed at:
Infinity Imaging Systems
Delhi

Preface

This book "Text Book of Heat" primarily intended for students preparing for degree and honours students of various Indian universities. Since the present day student is some what pressed of time the treatment has been kept short and direct only such historical and other additional information has been given as may possible interest the more serious type of student. An attempt has been made to make the language as simple as possible.

We hope this book will be found useful by the students and teacher in the various universities and institutions of India.

Author

Contents

Pages

Preface

1. **Heat and Tcmperature** 1

Concept of Heat and Temperature, Thermometry, Types of Thermometers, Centigrade and Fahrenheit Scales, Relation Between Celsius. Kelvin, Fahrenheit and Rankine Scales of Temperature, Liquid Thermometer, Errors and Corrections in a Mercury Thermometer, Gas Equation, Advantages of a Gas Thermometer, Callendar's Compensated Constant Prefigure Air Thermometer, Jolly's Constant Volume Air Thermometer, Constant Volume Hydrogen Thermometer, Gas Thermometer Corrections, Second Correction (Perfect or Ideal Gas Scale), Platinum Resistance Thermometer, Seeback Effect, Thermo-Electric Thermometer, Helium Vapour Pressure Thermometer, Standardization and Temperature Scale, Absolute Zero and Ice Point, Low Temperature Measurement, High Temperature Measurement.

2. **Thermal Expansion** 49

General, *Section I: Expansion of Solids*, Linear Expansion, Surface or Supercial Expansion, Cubical Expansion, Relation Between the Three Coefficients of Expansion, Relation Between Cubical Expansion and Density, Experimental Determination of the Coefficient of Linear Expansion of a Solid, Other Modifications to Fizeau's Method, Variation of the Coefficient of Expansion with Temperature, Gruneisen's Law, Linear Coefficient of Expansion of Crystals by X-ray Methods, The Coefficient of Cubical Expansion of a Crystal, Coefficient of Linear Expansion in a Direction Other Than the Principal Axes, Solids with Low Expansion Coefficient, Some Practical Applications of Thermal Expansion, *Section II: Expansion of Liquids*, Volume Expansion Coefficients, Apparent and Real Expansion of a Liquid, Relation Between γ_a and γ_r, Relation Between Volume and Density of a Liquid, Determination of Expansion Coefficients of Liquids, Expansion of Water, Hope's Experiment, Joule and Plafair's Experiment, Other Cases of Anomalous Expansion, Units of Mass and Volume, Practical Applications of Liquid Expansion, *Section III: Expansion of Gases*, Volume and Pressure Coefficients of a Gas, Determination of Volume Coefficient of a Gas, Determination of Pressure Coefficient of a Gas, Equality of Volume and Pressure Coefficients, Absolute Zero and Absolute Scale of Temperature, Application of Mansion of Gases,

3. Calorimetry 120

Definitions, Regnault's Method of Mixtures, Copper Block Calorimeter, Nernst Vacuum Calorimeter, Newton's Law of Cooling, Specific Heat of a Liquid—Joules Electrical Method, Specific Heat of a Liquid—Calorimeter and Barnes' Continuous Flow Method, Experimental Determination of Heat Capacities, Adiabatic Vacuum Calorimeter, Two Specific Heats of a Gas, Specific Heat of a Gas at Constant Volume—Jolly's Differential Steam Calorimeter, Specific Heat of a Gas at Constant Pressure— (Regnault's Method), Continuous Flow Electrical Method, Specific Heat of a Gas at Low Temperatures, Calorific Value of Fuels, Bell Calorimeter, Dulong and Petit's Law, Variation of Specific Heat and Atomic Heat with Temperature, Quantum Theory,

4. Change of State 161

Introduction, Latent of Heat of Fusion, Laws of Fusion, Practical Applications, Effect of Pressure on Freezing Point of Ice—Revelation, Impurities Lower Freezing Point, Determination of Melting Point of Wax, Determination of Latent Heat of Fusion of Ice, Heat Absorbed in Solution, Vaporization and Condensation, Laws of Ebullition or Boiling,

Change in Boiling Point with Pressure, Franklin's Experiment, Latent Heat of Vaporization, Bunsen's Ice Calorimeter, Cooling Effect due to Vaporization, Cryophorus, Examples of Cooling Caved by Evaporation, Refrigeration, Ammonia Ice Plant, Solid Carbon Dioxide—(Dry Ice), Gas and Vapour, Saturated and Unsaturated Vapours, Vapour Pressure of Liquids, Triple Point, Gibbs' Phase Rule, Hygrometry, Deniell's Hygrometer, Regnault's Hygrometer, Wet and Dry Bulb Hygrometer, Weather,

5. Nature of Heat **202**

Introduction, Caloric Theory of Heat, Failure of Caloric Theory, Mechanical Equivalent of Heat (Joules Experiments), Rowland's Experiment, Searle's Friction Cone Method, Joule's Calorimeter (Determination of J), Callendar and Barnes Continuous Flow Method (Determination of J), Jaegar and Steinwehr's Method, Kinetic Theory of Matter, The Three States of Matter, Concept of Ideal or Perfect Gas, Kinetic Theory of Gases, Expression for the Pressure of a Gas, Kinetic Energy Per Unit Volume of a Gas, Kinetic Interpretation of Temperature, Derivation of Gas Equation, Derivation of Gas Laws, Avogadro's Hypothesis, Graham's Law of Diffusion of Gases, Degrees of Freedom and Maxwell's Law of Equipartition of Energy, Atomicity of Gases, Maxwell's Law of Distribution of Velocity, Experimental Verification of Velocity Distribution, Mean Free Path, Transport Phenomena, Viscosity of Gases, Thermal Conductivity of Gases, Atomic Heat of Solids, Change of State, Continuity of State, Andrew's Experiments of Carbon-Dioxide, Amagat's Experiment, Holborn's Experiment, Behaviour of Gases at High Pressure, Van der Waals Equation of State, Critical Constants, Corresponding States, Coefficient of Van der Waals Constants, Reduced Equation of State, Properties of Matter Near Critical Point, Experimental Determination of Critical Constants, Intermolecular Attraction, Porous Plug Experiment, Theory or Porous Plug Experiment, Joule-Kelvin Effect—Temperature of Inversion, Relation Between Boyle Temperature, Temperature of Inversion and Critical Temperature,

6. Liquefaction of Gases **298**

Introduction, Cascade Process— Liquefaction of Oxygen, Liquefaction of Air—Linde's Process, Liquefaction of Hydrogen, Solidification of Hydrogen, Claude's Process—Liquefaction of Air, Liquefaction of Helium—K. Onnes Method, Helium I and Helium II, Production of Low Temperatures, Adiabatic Demagnetisation, Conversion of Magnetic Temperature to Kelvin Temperature, Helium Vapour Pressure Thermometer, Super Conductivity, Electrolux Refrigerator.

1

Heat and Temperature

1.1 CONCEPT OF HEAT AND TEMPERATURE

We may regard heat as the physical cause of the sensation of hotness, coldness and temperature as the degree of hotness or coldness. We know that a liquid flows only from a higher to a lower level until a common level is attained, irrespectable we the quantity of liquid at two levels. It we replace quantity of heat liquid by quantity of heat and its level of temperature. We have heat flow from a body at higher temperature to the one at a lower temperature irrespective of the quantities of heat in the two bodies. This heat flow continuous until the two bodies attain a common temperature. Therefore, heat can be defined as energy in transit.

Temperature of a system can be defined as the property that determines whether or not the body is in thermal equilibrium with the neighbouring systems. If a number of systems are in thermal equilibrium, this common property of the system can be represented by a single numerical value called the temperature. It means that if two systems are not in thermal equilibrium, they are at different temperatures.

Measurement of temperature of a body accurately is one of the important branches of heat in physics. It also becomes necessary to measure high temperatures and low temperatures. To make this measurement possible, it is necessary to construct a suitable scale of temperature. The scale chosen must be precise and consistent and the temperatures measured on this scale must be accurate. Assessing the temperature of a body by mere sense of touch or comparing the degree of hotness of a body with respect to another body does not help in measuring the temperature quantitatively and accurately.

1.2 THERMOMETRY

The branch of heat relating to the measurement of temperature of a body is called *thermometry*. Thermometer is an instrument used to measure the temperature of a body.

The essential requisites of a thermometer are given as under:

(1) Construction,

(2) Calibration, and

(3) Sensitiveness.

For the construction of a thermometer, the proper choice of a substance, whose physical property varies uniformly with rise in temperature, is essential.

1. Construction

The physical property of a substance plays an important role in the construction of a thermometer. In a mercury thermometer, the principle of expansion of mercury with rise in temperature is used. The platinum resistance thermometer is based on the principle of the change in resistance with change in temperature. The gas thermometer is based on the principle of change in volume or pressure with change in temperature.

2. Calibration

When a thermometer is constructed, it should be properly calibrated. The standard fixed points are selected for calibrating a thermometer. Melting point of ice, boiling point of water, melting point of silver and melting point of gold are taken as fixed points. The scales are built by dividing the interval between the two fixed points into equal parts. Centigrade scale is built by dividing the interval between the melting point of ice and the boiling point of water (under normal pressure) into 100 equal parts and each part represents 1°C. Similarly, Fahrenheit scale is built by dividing this interval into 180 equal parts.

3. Sensitiveness

The instrument, once constructed and calibrated, should also be sensitive. The thermometer will be sensitive if:

(i) it can detect even small changes in temperature,

(ii) it shows the temperature of a body in a short time and

(iii) it does not take large quantity of heal for its own heating from the body whose temperature is being measured.

Fixed Points	Degrees Celsius
Boiling point of oxygen	–182.97
Ice point	0.00
Steam point	100.00
Boiling point of sulphur	444.60
Melting point of silver	960.80
Melting point of gold	1063.00
Melting point of cobalt	1492.00
Melting point of platinum	1769.00
Melting point of rhodium	1960.00
Melting point of tungsten	3380.00

1.3 TYPES OF THERMOMETERS

There are different kinds of thermometers:

1. Liquid Thermometers

These thermometers are based on the principle of change in volume of a liquid with change in temperature. Mercury and alcohol thermometers are based on this principle.

2. Gas Thermometers

These are based on the principle of change in pressure or volume with change in temperature, *e.g.*, Callendar's constant pressure thermometer, constant volume hydrogen thermometer etc.

3. Resistance Thermometers

These are based on the principle of change in resistance with change in temperature, *e.g.*, platinum resistance thermometer.

4. Thermoelectric Thermometers

These are based on the principle of thermoelectricity, *i.e.*, production of thermo-E.M.F. in a thermo-couple when the two junctions are at different temperatures.

The various thermocouples commonly used are:

(1) Copper and constantan

(2) Iror and constantan

(3) Chromel and constantan

(4) Chromel and alumel

(5) Platinum and Rhodium.

5. Radiation Thermometers

These are based on the quantity of heat radiations emitted by a body *e.g.*, furnaces. These instruments are known as pyrometer.

6. Vapour Pressure Thermometers

These are based on the principle of change of vapour pressure with change in temperature. These are used to measure low temperatures, *e.g.*, helium vapour pressure thermometer etc.

7. Bimetallic Thermometers

These thermometers are based on the principle of expansion of solids. A bimetallic strip is taken in the form of a spiral. Its one end is fixed and the other end is attached to a long pointer. The pointer moves on a scale, calibrated in degrees. These thermometers are used in meteorology for recording the changes in temperature during the day. They are also used to measure temperatures at high altitudes.

8. Magnetic Thermometers

These thermometers are based on the principle of change in the susceptibility of a substance with temperature. These thermometers are useful for measuring low temperatures near the absolute zero temperature.

1.4 CENTIGRADE AND FAHRENHEIT SCALES

The earliest thermometer was constructed by Galileo in 1593. Newton suggested the necessity of the fixed points. The temperature of the melting point of ice is taken as the lower fixed point and the temperature of steam at a pressure of 76 cm of Hg (normal pressure) is taken as the upper fixed point.

Centigrade (or Celsius) Scale : Celsius, in 1742, suggested the centigrade system of temperature. He marked zero at the lower fixed point and 100 at the upper fixed point. The interval between the two fixed points is divided into 100 equal parts. Each part or degree represents 1°C or 1° Celsius. The scale is also known as Celsius scale.

Fahrenheit Scale : Fahrenheit, in 1720, suggested this scale by taking zero as the temperature of the freezing mixture. It appears that he took 100 degrees as the temperature of the human body. Later the correct temperature of the human body on this scale was found to be 98-4°F. The lower fixed point is marked as 32 and the upper fixed point is marked as 212. The interval is divided into 180 equal parts. Each part or degree represents 1°F.

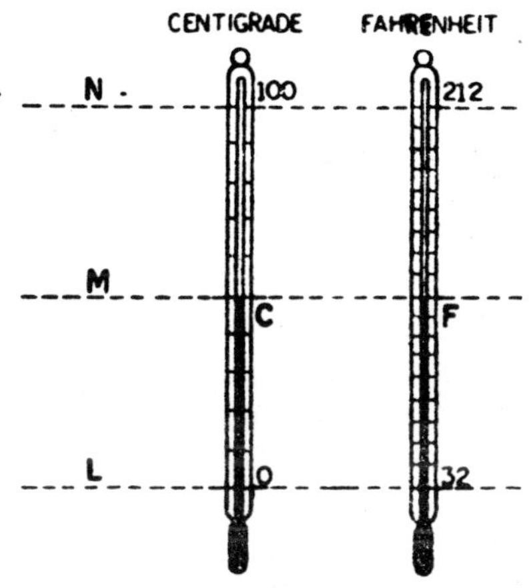

Fig. 1.1

Relation : Consider two identical thermometers marked in Centigrade and Fahrenheit scales. Place the two thermometers in a bath at a certain fixed temperature. Mercury in each thermometer stands to the same level At (Fig. 1.1).

$$\therefore \quad \frac{ML}{NL} = \frac{C - 0}{100 - 0}$$

$$= \frac{F - 32}{212 - 32}$$

or $$\frac{C}{100} = \frac{F - 32}{180}$$

1.5 RELATION BETWEEN CELSIUS, KELVIN, FAHRENHEIT AND RANKINE SCALES OF TEMPERATURE

In Fig. 1.2, the temperatures of the absolute zero, the melting point of ice and the boiling point of water as measured on the Celsius (Centigrade), Kelvin (absolute), Fahrenheit and Rankine scales are shown.

Celsius and Fahrenheit scales show the same reading at –40° *i.e.*, –40°C = –40°F. The Kelvin scale and the Rankine scale agree at zero degree only.

The temperatures represented in (Fig. 1.2) have been rounded off to the nearest degree.

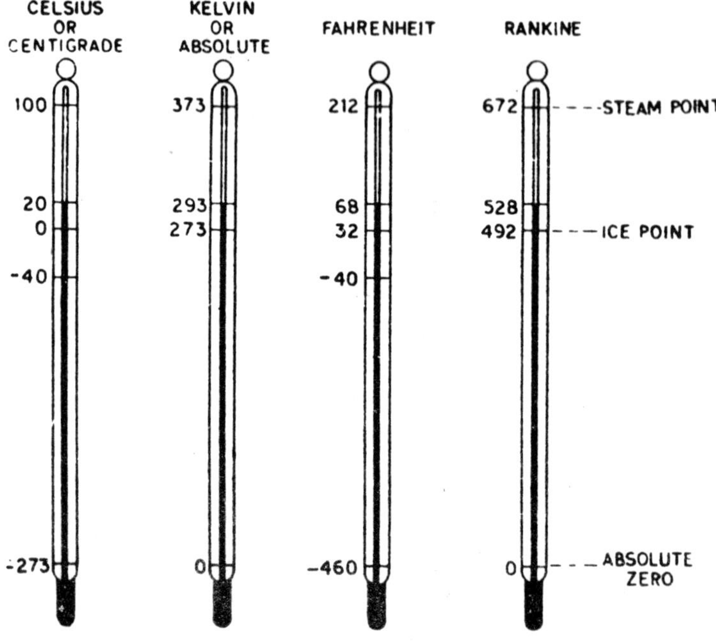

Fig. 1.2

Relation :

$$\frac{C - 0}{100} = \frac{F - 32}{180}$$

$$= \frac{K - 273}{100} = \frac{R - 492}{180}$$

Also $$\frac{K}{100} = \frac{R}{180}$$

Example 1:

The temperature of the surface of the sun is about 6500°C. What is this temperature (i) on the Rankine scale and (ii) on the Kelvin scale?

Solution:

We have $$\frac{C - 0}{100} = \frac{K - 273}{100} = \frac{R - 492}{180}$$

Here, $C = 6500°C$

$$\frac{6500}{100} = \frac{K - 273}{100} = \frac{R - 492}{180}$$

$$K = 6500 + 273 = 6773 \text{ K}$$

$$R = 12{,}192°R$$

The temperature of the surface of the sun corresponding to 6500°C is (i) 6773 K and (it) 12,192°R.

Example 2:

The normal boiling point of liquid oxygen is –183°C. What is this temperature on (i) Kelvin scale and (ii) Rankine scale?

Solution:

We have $$\frac{C - 0}{100} = \frac{K - 273}{100} = \frac{R - 492}{180}$$

Here, $C = -183°C$

$$\therefore \quad \frac{-183}{100} = \frac{K - 273}{100} = \frac{R - 492}{180}$$

$$K = 90°K$$

$$R = 162.60°R$$

The boiling point of liquid oxygen corresponding to –183°C is (i) 90 K and (ii) 162-6°R.

Example 3:

At what temperature do the Kelvin and Fahrenheit scales coincide?

Solution:

We have $\frac{K - 273}{100} = \frac{F - 32}{180}$

$$\frac{x - 273}{100} = \frac{x - 32}{180}$$

$$x = 574.25°$$

$$574.25\ K = 574.25°F$$

Example 4:

At what temperature do the Celsius and the Fahrenheit scales coincide?

Solution:

We have $\frac{C - 0}{100} = \frac{F - 32}{180}$

$$\frac{x - 0}{100} = \frac{x - 32}{180}$$

$$x = -40°$$

$$\therefore \quad -40°C = 40°F$$

Example 5:

The boiling point of liquid hydrogen is 20.2 K. Convert this temperature into degree Rankine.

Solution:

We have $\frac{K - 273}{100} = \frac{R - 492}{180}$

Here $K = 20.2°\ K$

$$\therefore \quad \frac{20.2 - 273}{100} = \frac{R - 492}{180}$$

$$R = 36.96°R$$

$$\therefore \quad 20.2\ K = 36.96°R.$$

1.6 LIQUID THERMOMETER

Liquid thermometers are based on the principle of change of volume of the liquid with change in temperature. These thermometers are suitable for a narrow range of measurement. The most commonly used liquids are (i) mercury and (ii) alcohol. The range of mercury thermometer is –39°C to 357°C. Alcohol thermometers are used only to measure temperature near ice point.

Mercury Thermometer

Mercury is usually selected for use in liquid in glass thermometers for the following reasons:

(1) It has a low specific heat and hence it absorbs little heat from the body whose temperature is being measured.

(2) It is a good conductor of heat and takes the temperature of the body quickly.

(3) It can be easily seen in a fine capillary tube and the thermometer can be made sensitive.

(4) It does not wet the wall of the glass tube. This is an important point for the construction of a thermometer.

(5) It has a uniform coefficient of expansion over a wide range of temperature.

(6) It remains a liquid over a large range. Its freezing point is –39°C and the boiling point is 357°C.

1.7 ERRORS AND CORRECTIONS IN A MERCURY THERMOMETER

1. Change of Zero

When a thermometer is graduated after sealing, there will be shift in the zero mark. This is due to the fact that when a thermometer is subjected to a high temperature during construction, glass takes a long time for contraction. Consequently, when the thermometer is placed in melting ice, the level of mercury will be above the zero mark (say at + 0.5°C). To correct the error, 0.5°C is subtracted from the observed reading. While manufacturing, the scaled thermometer tubes are allowed sufficient time and are graduated after a number of years.

2. Recent Heating and Cooling

When a thermometer is used to measure a very high temperature, the thermometer bulb expands. If the same thermometer is immediately used to measure low temperature, the level of mercury will be lower than the correct reading. Suppose the thermometer when placed in melting ice, reads –0.3°C, the correction to be applied is +0.3°C.

3. Due to Exposed Stem

While calibrating a thermometer, the thermometer as a whole is kept in steam. While measuring the temperature, only the bulb is immersed in the bath whose temperature is to be measured. Due to this reason the observed reading is lower than the correct reading. To correct this error, maximum possible length of the stem should be immersed in the bath whose temperature is to be measured.

4. Inequality of the Bore

The thermometer is calibrated after marking the lower and the upper fixed points. The distance between the two fixed marks is divided into equal parts assuming that the bore of the thermometer is uniform. If the bore is not uniform, at positions where the bore has lesser diameter, the observed reading is higher than the actual reading. Similarly, when the bore has a higher diameter, the observed reading is less. To correct this error the capillary tube is initially calibrated and a graph is provided which gives the correct readings against the observed readings.

5. Position and Pressure

While graduating, if the thermometer was held in the horizontal position and while taking the temperature it is kept in the vertical position, the observed readings will be lower than the correct readings. This is due to the pressure of mercury column. To avoid this error, the thermometer is used in the same position in which it was calibrated.

6. Surface Tension

Mercury is depressed in a capillary tube because it does not wet the glass tube. The depression is different at different positions of the stem if the bore is not uniform. To correct this error, the reading of the thermometer is taken with a rising mercury column.

7. Thermal Capacity

If the thermal capacity of the bulb is large, a part of the heat from the bath is used in heating the bulb. Therefore, the thermometer reads a lower temperature. To correct this error, small thermometer bulbs with low thermal capacity are used.

8. Effect of the Bulb

In a mercury thermometer the glass bulb has an insulating effect. Due to this reason the mercury inside the bulb may not attain the temperature of the bath.

1.8 GAS EQUATION

Consider a perfect gas at a pressure P, temperature T and volume V.

Let the pressure of the gas change from P to P_1 at constant temperature T such that the volume changes from V to v.

Applying Boyle's law,

$$P_1 v = PV$$

or
$$v = \frac{PV}{P_1} \qquad \text{...(i)}$$

Now, let the temperature of the gas change from T to T_1 at constant pressure.

Initial Pressure = P, Volume = v, Temperature = T

Final Pressure = P_1 Volume = V_1, Temperature = T_1

Applying Charles' law,

$$\therefore \quad \frac{v}{T} = \frac{V_1}{T_1} \qquad \text{...(ii)}$$

Substituting the value of v in equation (ii)

$$\frac{PV}{TP_1} = \frac{V_1}{T_1}$$

or
$$\frac{PV}{T} = \frac{P_1 V_1}{T_1}$$

$$\therefore \qquad \frac{PV}{T} = \text{constant.} \qquad ...(iii)$$

Universal gas Constant

When one gram molecule of a gas at N.T.P. is taken, the constant in equation {Hi) is S and is called '*universal gas constant*'. Its value is the same for all gases.

$$\frac{PV}{T} = R$$

$$\text{or} \qquad PV = RT \qquad ...(i)$$

One gram molecule of a gas at N.T.P. has a volume = 22,400 cc

At normal pressure

$$P = 76 \times 13.6 \times 980 \text{ dynes/sq cm}$$

$$V = 22{,}400 \text{ cc}$$

$$T = 273 \text{ K}$$

$$R = \frac{76 \times 13.6 \times 980 \times 22.400}{273}$$

$$= 8.31 \times 10^7 \text{ ergs/g-mol-K}$$

$$R = \frac{8.31 \times 10^7}{4.2 \times 10^7} = 1.986 \text{ cal/g} - \text{mol} - \text{K}$$

For n gram molecules of a gas, the gas equation is

$$PV = nRT. \qquad ...(ii)$$

Ordinary gas constant. When one gram of a gas at N.T.P. is taken, the gas constant is r and is known as ordinary gas constant. Its value is different for different gases.

$$\text{Here} \qquad \frac{PV}{T} = r$$

$$\text{or} \qquad PV = rT.$$

(i) For Oxygen: One gram of oxygen at N.T.P. has a volume

$$= \frac{22{,}400}{32} \text{ cc}$$

$$r = \frac{PV}{T} = \frac{76 \times 13.6 \times 980 \times 22,400}{273 \times 32}$$

$$= \left(\frac{8.31}{32}\right) \times 10^7 \text{ ergs/g} - K$$

(ii) For Hydrogen: One gram of hydrogen at N.T.P. has a volume

$$= \frac{22,400}{2} \text{ cc}$$

$$r = \frac{PV}{T} = \frac{76 \times 13.6 \times 980 \times 22,400}{273 \times 2}$$

$$= \left(\frac{8.31}{2}\right) \times 10^7 \text{ ergs/g} - K$$

Similarly r for nitrogen

$$= \left(\frac{8.31}{28}\right) \times 10^7 \text{ ergs/g} - K$$

and $\quad r \text{ for } CO_2 = \left(\frac{8.31}{44}\right) \times 10^7 \text{ ergs/g} - K$

In general, $r = \dfrac{R}{\text{Molecular wt. of the gas in grams}}$

1.9 ADVANTAGES OF A GAS THERMOMETER

1. The coefficient of expansion of gases is very large as compared to liquids. Therefore, gas thermometers are sensitive.
2. The coefficient and the rate of expansion of all gases are the same under similar conditions.
3. The coefficient of expansion of the material of the bulb of the thermometer is negligible in comparison to the coefficient of expansion of a gas.
4. The gases expand uniformly and regularly over a wide range of temperature.
5. The thermal capacity of a gas is low as compared to liquids. Hence even small changes of temperature can be recorded accurately.
6. Gases can be obtained in a pure form.

7. Gas thermometers can be used over a wide range of temperature. They are suitable to measure high and low temperatures.
8. The temperatures measured with a gas thermometer agree with the temperatures on the thermodynamic scale.

Gas thermometers are not suitable for routine work. They are large and cumbersome. They can be used only in one position. They are mainly used to standardize and calibrate other thermometers.

1.10 CALLENDAR'S COMPENSATED CONSTANT PREFIGURE AIR THERMOMETER

It is based on the principle that, pressure remaining constant, the volume of a given mass of gas varies directly as its absolute temperature. It consists of a silica bulb A connected to the reservoir

R containing mercury (Fig. 1.3). *EF* is the connecting tube. The compensating bulb B is exactly similar to the silica bulb A. C_1C_2 is the compensating tube having a volume equal to the tube EF.

Initially, the reservoir R is filled with mercury up to the zero mark and the stop-cock is closed. The bulbs A, B and S are immersed in melting ice. The tubes are sealed when the pressure on the two sides as shown by the manometer is the same. Therefore, pressure of air in A and B is the same.

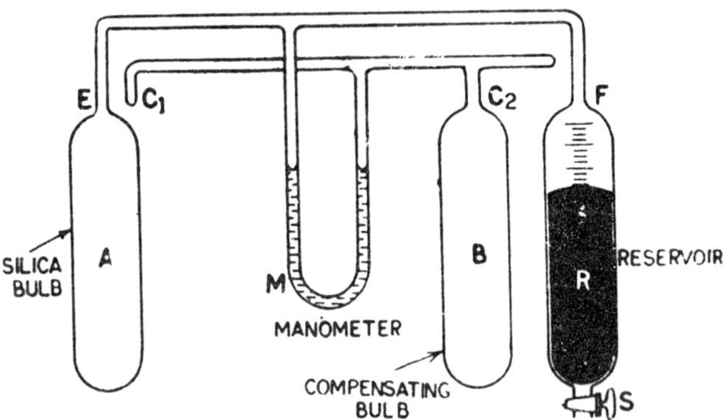

Fig. 1.3

The manometer M contains sulphuric acid. When the pressure on the two sides of the manometer is the same, the levels of sulphuric acid in the two limbs will be the same. In this way the pressure of the gas or air can be maintained constant at ail temperatures.

Suppose, the bulb A and the connecting tube *EF* contain n gram molecules of air. Also the bulb B and the compensating tube C_1C_2 contain n gram molecules of air. The bulb A is immersed in the bath whose temperature is to be measured. The bulb B is kept in melting ice. Tubes *EF* and C_1C_2 are at the room temperature. The air in the bulb A attains the temperature of the bath. Pressure of air in A increases and the manometer shows the difference in level. Mercury is allowed to flow out of the reservoir R through the stop-cock S and the stop-cock is closed just at the moment when the pressure on the two sides of the manometer is the same.

Volume of the bulb A = Volume of the bulb B = V

Volume of the connecting tube *EF* = Volume of the compensating tube C_1C_2 = x

Volume of air in the reservoir R at the end = v

Pressure of air on each side = P

Temperature of the bath = T

Room temperature = T

Temperature of melting ice = T_0

For the thermometric part,

$$\frac{PV}{T} + \frac{Px}{T_1} + \frac{Pv}{T_0} = nR \qquad ...(i)$$

For the compensating part

$$\frac{PV}{T_0} + \frac{Px}{T_1} = nR \qquad ...(ii)$$

Equating (i) and (ii)

$$\frac{PV}{T} + \frac{Px}{T_1} + \frac{Pv}{T_0} = \frac{PV}{T_0} + \frac{Px}{T_1}$$

$$\frac{V}{T} + \frac{V - v}{T_0}$$

$$T = \left[\frac{V}{V - v}\right] T_0 = \left[\frac{V}{V - v}\right] 273 \qquad ...(iii)$$

Here T will be obtained in degrees Kelvin

Let the temperature of the bath be t°C

Then, $\quad T = (273 + t)$

$$\therefore \quad 273 + t\left(\frac{V}{V - v}\right)273$$

$$t = \left[\left(\frac{v}{V - v}\right)273\right] °C \qquad \text{...(iv)}$$

This thermometer is useful to measure temperatures upto 600°C. However, the results obtained with this thermometer are not very accurate.

Example 6:

The bulb of the Callendar's compensated constant pressure air thermometer is 1,000 cm^3. When the bulb is immersed in a bath, 1M cm^3 of mercury has to be drawn out of the reservoir. Calculate the temperature of the bath on the Celsius scale.

Solution:

We have V = 1,000 cm^3

Here $\quad$ 7 = 1,000 cm^3

$$t = \left(\frac{v}{V - v}\right)273$$

$$= \left(\frac{100}{1000 - 100}\right)273$$

$$= 30.33°C.$$

1.11 JOLLY'S CONSTANT VOLUME AIR THERMOMETER

It consists of a glass bulb B connected to a glass tubing. The end of the glass tube is connected to the reservoir of mercury through a rubber tubing. M is a fixed mark on the glass tube. The difference in the levels of mercury in R and M is observed on the scale 8. The bulb B is filled with 1/7 of its volume with mercury so that the expansion of the bulb B is compensated by the expansion of mercury in the bulb. This keeps the volume of air in the bulb up to the mark M constant (Fig. 1.4).

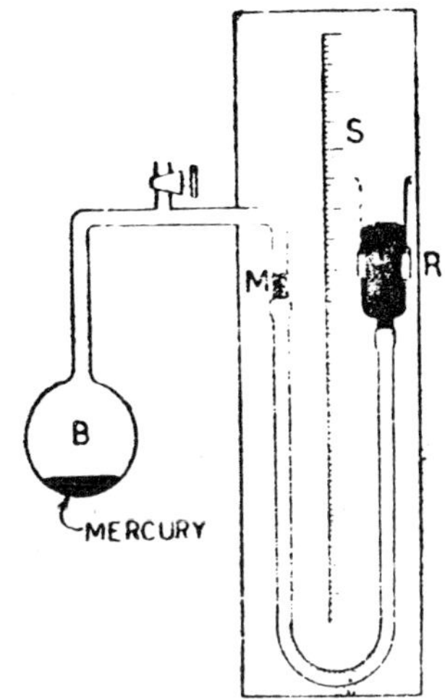

Fig. 1.4

Working : (1) The bulb B is kept in melting ice. The reservoir S is suitably adjusted so that the level of mercury stands at M. Suppose the difference in level between R and M is h_0. If P is the atmospheric pressure, then

$$P_0 = P + h_0 \qquad ...(i)$$

(2) The bulb B is kept in boiling water. The reservoir K is adjusted so that the level of mercury is at M. Let Also be the difference in levels.

$$P_{100} = P + h_{100} \qquad ...(ii)$$

(3) The bulb B is kept in the bath whose temperature is to be measured. The reservoir B is adjusted so that the level of mercury is at M. Let h_t be the difference in levels.

$$\therefore \quad P_t = P + h_t \qquad ...(iii)$$

Volume remaining constant, the pressure increases according to the relation

$$P_t = P_0 (1 + \gamma^t)$$

Also $\quad P_{100} = P_0 (1 + \gamma 100)$

or $\quad P_t - P_0 = P_0 \gamma t \qquad ...(iv)$

and $\quad P_{100} - P_0 = P_0 \gamma 100 \qquad ...(v)$

Dividing (iv) by (v)

$$t = \left(\frac{P_t - P_0}{P_{100} - P_0}\right) \times 100 \qquad \text{...(vi)}$$

$$t = \left(\frac{h_t - h_0}{h_{100} - h_0}\right) \times 100 \qquad \text{...(vii)}$$

Hence t can be calculated.

1.12 CONSTANT VOLUME HYDROGEN THERMOMETER

It was first designed try Harker and Chappius and is known as the International standard thermometer. It is based on the principle that when the volume is kept constant, the pressure of a gas increases with rise in temperature according to the relation

$$Pt = P_0(1 + \gamma t)$$

It consists of a platinum-iridium bulb B connected to the tube A. The reservoir K containing mercury is connected to the tubes C and A as shown in Fig. 1.5. A movable barometric tube T with a bulb D is clamped to a stand. It can be moved vertically up or down as desired. P_1 and P_2 are the two ivory tips in the same vertical line. Above the level of mercury in D, there is Toricellian vacuum (Fig. 1.5).

Working : Initially the bulb is filled with pure dry hydrogen at a pressure higher than the atmospheric pressure.

(1) The bulb B is kept in melting ice. The reservoir S is adjusted so that the level of mercury in A just touches the tip of the ivory peg P_1. The tube T is adjusted so that the level of mercury in D just touches the tip of the ivory peg P_2. Suppose the difference in levels of mercury in A and C is h_0 and the atmospheric pressure is P.

    Then $P_0 = P + h_0$ The difference in levels of mercury in C and D is equal to the atmospheric pressure P. It means the distance between the tip of P_1 and P_2 as measured on the scale (or with the help of a cathetometer) gives P_0 directly.

(2) The bulb B is kept in steam or boiling water. R and T are adjusted so that the level of mercury in A just touches the tip of P_1 and that in D just touches the tip of P_2. The difference in the levels of P_1 and P_2 gives P_{100}.

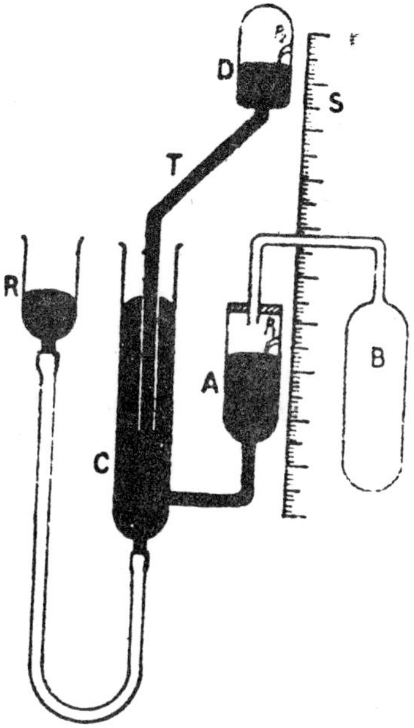

Fig. 1.5

(3) Place the bulb B in the bath whose temperature is to be measured. R and T are adjusted in the same way and the difference in levels of P_1 and P_2 gives P_t.

Here $P_t = P_0 (1 + \gamma t)$

$$P_{100} = P_0 (1 + \gamma 100)$$

$\therefore$ $P_t - P_0 + P_0\gamma t$ ...(i)

$$P_{100} - P_0 = P_0\gamma\ 100 \qquad \text{...(ii)}$$

Dividing (i) by (ii),

$$t = \left(\frac{P_t - P_0}{P_{100} - P_0} \right) \times 100 \qquad \text{...(iii)}$$

The constant volume hydrogen thermometer can be used over a wide range of temperature. With hydrogen and platinum-iridium bulb, the range is –200°C to 500°C.

Beyond 500°C, hydrogen diffuses through platinum. With a porcelain bulb temperatures up to 110°C can be measured. Beyond 1100°C hydrogen creeps along the wall of the bulb. So it is replaced by nitrogen for measuring temperatures up to 1500°C. Below –200°C hydrogen is replaced by helium.

1.13 GAS THERMOMETER CORRECTIONS

While discussing the constant volume hydrogen thermometer, the following assumptions were made:

(i) The whole gas attains the temperature to be measured and that the volume of the gas remains constant.

(ii) The gas obeys the perfect gas laws.

In practice, however, these two assumptions are not true and corrections are to be applied.

First correction : The main sources of error in a constant volume gas thermometer are:

(i) The gas in the dead space is not at the same temperature as the gas in the bulb. The dead space consists of the space inside the capillary and in the manometer between the mercury level and the index.

(ii) There is increase in the volume of the bulb due to rise in temperature.

(iii) There is a change in the volume of the bulb due to change in pressure.

Let the volume of the bulb at 0°C be V_0 and P_0 the pressure of the gas at 0°C. Suppose the volume of the dead space is v and the temperature of the dead space throughout the experiment is θ°C. The reduced volume of the whole gas at N.T.P.

$$= \left(V_0 + \frac{v}{1 + \gamma\theta} \right) \frac{P_0}{76} \qquad ...(i)$$

Here γ is the coefficient of expansion of the gas,

When the bulb is heated to t °C, the pressure of the gas becomes $(P_0 + h)$.

The reduced volume of the gas at N.T.P. will be,

$$\left\{\left(\frac{V_0(1+\gamma_g^t}{1+\gamma t}\right)+\frac{v}{1+\gamma\theta}\right\}\left(\frac{P_0+h}{76}\right) \quad \text{...(ii)}$$

Here, γ_g is the volume coefficient of expansion of the bulb and V is the volume coefficient of expansion of the bulb due to internal pressure.

Equations (i) and (ii) represent the reduced volume of the gas at NTP at 0°C and t °C. As the mass of the enclosed gas is constant, these two volumes must be equal.

Equating (i) and (ii) we get,

$$V_0P_0 = \frac{\left\{V_0\,(1+\gamma_g^t)+bh\right\}(P_0+h)}{1+\gamma t}+\frac{vh}{1+\gamma\theta} \quad \text{...(iii)}$$

$$\text{or } \gamma t = \left(\frac{P_0+h}{P_0}\right)\left(\frac{V_0(1+\gamma_g^t)\;bh}{V_0}\right)+\frac{hv}{P_0V_0}\left(\frac{1+\gamma t}{1+\gamma\theta}\right)-1 \quad \text{...(iv)}$$

Here t is the temperature of the gas and it occurs on both sides of the equation (iv). In case the above corrections are not taken into account, equation (iv) reduces to,

$$\gamma t = \frac{h}{P_0}$$

$$\text{or} \qquad t = \frac{h}{\gamma P_0} \quad \text{...(v)}$$

Initially, the value of t is calculated from equation (v) and this approximate value is substituted on the right hand side of equation (iv). After substituting, a new value of t is obtained for the left hand side. This new value is again substituted on the right hand side and the value of t on the left hand side is once again calculated. This method of successive substitution is performed two or three times depending upon the accuracy desired.

1.14 SECOND CORRECTION (PERFECT OR IDEAL GAS SCALE)

All practical gases *viz*., hydrogen, oxygen, nitrogen etc., show deviations from Boyle's law at high pressures and low temperatures. Hence the temperatures obtained with a constant volume or a constant pressure gas thermometer have to be corrected so as to compensate for the departure from the ideal gas. On the constant volume scale, the temperature t_v is written as,

$$t_v = \left[\frac{P_t - P_t}{P_{100} - P_0}\right] \times 100 \qquad ...(i)$$

Similarly on the constant pressure scale the temperature t_p is written as

$$t_p = \left(\frac{V_t - V_0}{V_{100} - V_0}\right) \times 100 \qquad ...(ii)$$

If the mass of the gas enclosed is the same in both the thermometers, then at 0°C, both will have the same pressure P_0 and volume V_0. In the case of constant volume gas thermometer, when the temperature is increased to t °C, the pressure becomes P_t. In the constant pressure gas thermometer, at t °C, the volume is V_t. It the gas is ideal, in that case

$$V_0P_t = (PV)_t = V_tP_0 \qquad ...(iii)$$

Multiplying and dividing the right hand side of equation (i) by V_0,

$$t_v = \left[\frac{(P_t - P_0)V_0}{(P_{100} - P_0)V_0}\right] \times 100$$

$$t_v = \left[\frac{P_tV_0 - P_0V_0}{P_{100}V_0 - P_0V_0}\right] \times 100 \qquad ...(iv)$$

But from equation (iii),

$$P_tV_0 = (PV)_t$$

and $$P_0V_0 = (PV)_0$$

$$P_{100}V_0 = (PV)_{100}$$

$$\therefore \quad t_v = \left[\frac{(PV)_t - (PV)_0}{(PV)_{100} - (PV)_0}\right] \times 100 \qquad ...(v)$$

Similarly it can be shown that

$$t_p = \left[\frac{(PV)_t - (PV)_0}{(PV)_{100} - (PV)_0}\right] \times 100 \qquad ...(vi)$$

Thus for an ideal gas, the temperatures measured on the constant pressure scale and constant volume scale are identical and both can be written as

$$t = \left[\frac{(PV)_t - (PV)_0}{(PV)_{100} - (PV)_0}\right] \times 100 \qquad ...(v)$$

But in actual practice no gas is ideal.

In the case of real gases, an empirical relation suggested by K-Onnes is

$$PV = A + BP + OP^2 + DP^2 + ... \qquad ...(vi)$$

Hew A B, C, D etc., are constants.

These constants are functions of temperature but are independent of pressure. The values of these coefficients are determined by performing compression experiments. These values for oxygen at 0°C are

$$A = 1.00130$$

$$B = -1.30143 \times 10^{-3}$$

$$C = 3.6898 \times 10^{-6}$$

Here P is measured in metres of Hg and when P = 1, V is also equal to one.

It can be seep that the constants A, B, C, D are of rapidly decreasing magnitude. In actual practice, for the measurement of temperature, constants C and D are usually negated. Therefore,

$$PV = A + BP \qquad ...(vii)$$

Using equation (vii), it is clear that when $f \to 0$, PV = A or the gas is ideal and obeys Boyle's Jaw.

In order to obtain the temperatures corresponding to an ideal gas, observations are made with a real gas and the values of the temperature, t_{lim} when $P \to 0$ are obtained. Here t_{lim} will correspond to the ideal gas temperature.

Constant Volume Scale

On the constant volume scale

$$t_v = \left(\frac{P_t - P_0}{P_{100} - P_0}\right) \times 100 \qquad ...(viii)$$

From the equation

PV = A + BP, we have

$$\left.\begin{aligned} P_0V_0 &= A_0 + B_0P_0 \\ P_{100}V_0 &= A_{100} + B_{100}P_{100} \\ P_tV_0 &= A_t + B_tP_t \end{aligned}\right\} \qquad \text{...(ix)}$$

Multiplying and dividing the right hand side of equation (xiii) by P_0,

$$t_v = \left(\frac{P_tV_0 - P_0V_0}{P_{100}V_0 - P_0V_0}\right) \times 100$$

$$\therefore\ t_v = \left[\frac{(A_t + B_tP_t) - (A_0 + B_0P_0)}{(A_{100} + B_{100}P_{100}) - (A_0 + B_0P_0)}\right] \times 100$$

$$t_v = \left[\frac{(A_t + A_0) + (B_tP_t - B_0P_0)}{(A_{100} - A_0) + (B_{100} - B_0P_0)}\right] \times 100$$

$$t_v = \left(\frac{A_t - A_0}{A_{100} - A_0}\right) \times 100 \left[\frac{\left(1 + \dfrac{B_tP_t - B_0P_0}{A_t - A_0}\right)}{\left(1 + \dfrac{B_{100}P_{100} - B_0P_0}{A_{100} - A_0}\right)}\right]$$

$$t_v = \left(\frac{A_t - A_0}{A_{100} - A_0}\right) \times 100 \left(1 + \frac{B_tP_t - B_0P_0}{A_t - A_0}\right) \times \left(1 - \frac{B_{100}P_{100} - B_0P_0}{A_{100} - A_0}\right)$$

$$t_v = \left(\frac{A_t - A_0}{A_{100} - A_0}\right) \times 100 \left(1 + \frac{B_tP_t - B_0P_0}{A_t - A_0} - \frac{B_{100}P_{100} - B_0P_0}{A_{100} - A_0}\right) \qquad \text{...(x)}$$

When $P \to 0$

$$t_{lim} = \left(\frac{A_t - A_0}{A_{100} - A_0}\right) \times 100$$

$$\therefore\ t_v = t_{lim}\left(1 + \frac{B_tP_t - B_0P_0}{A_t - A_0} - \frac{B_{100}P_{100} - B_0P_0}{A_{100} - A_0}\right) \qquad \text{...(xi)}$$

$$\text{or } t_{lim} - t_v = t_{lim}\left(\frac{B_{100}P_{100} - B_0P_0}{A_{100} - A_0} - \frac{B_tP_t - B_0P_0}{A_t - A_0}\right) \quad \text{...(xii)}$$

Equation (xii) gives the correction that has to be applied. The value of t_v is obtained with a thermometer and the value of the constants A_0, B_0, A_t, B_t, A_{100}, B_{100}, are obtained by actual compression experiments.

Initially the value of t_v is substituted for t_{lim}, on the right hand side of equation (xii) and the correction $t_{lim}-t_v$ is calculated. This gives a new value of t_{lim}. This is again substituted on the right hand side and this process of successive substitution is done two or three times. Finally t_{lim} gives the accuratc value of temperature on the *ideal gas scale.*

Constant Pressure Scale

On the constant pressure scale

$$t_p = \left(\frac{V_t - V_0}{V_{100} - V_0}\right) \times 100 \quad \text{...(xiii)}$$

From the equation

$$PV\ A + BP \text{ we have}$$

$$P_0V_0 = A_0P_0$$

$$P_0 = V_{100} = A_{100} + B_{100}P_0$$

$$P_0V_t = A_t + B_tP_0$$

Multiplying and dividing the right hand side of equation (xiii) by P_0,

$$t_p = \left(\frac{P_0V_t - P_0V_0}{P_0V_{100} - P_0V_0}\right) \times 100$$

$$t_p = \left[\frac{(A_t + B_tP_0) - (A_0 + B_0P_0)}{(A_{100} + B_{100}P_0) - (A_0 + B_0P_0)}\right] \times 100$$

$$t_p = \left(\frac{A_t - A_0}{A_{100} - A_0}\right) \times 100\left[1 + P_0\left(\frac{B_t - B_0}{A_t - A_0}\right) - P_0\left(\frac{B_{100} - B_0}{A_{100} - A_0}\right)\right]$$

When $P \rightarrow 0$

$$t_{lim} = \left(\frac{A_t - A_0}{A_{100} - A_0}\right) \times 100$$

$$\therefore \quad t_p = t_{lim}\left[1 + P_0\left(\frac{B_t - B_0}{A_t - A_0}\right) - P_0\left(\frac{B_{100} - B_0}{A_{100} - A_0}\right)\right]$$

$$\therefore \quad t_{lim} = t_p = t_{lim} P_0\left[\left(\frac{B_{100} - B_0}{A_{100} - A_0}\right) - \left(\frac{B_t - B_0}{A_t - A_0}\right)\right] \qquad \text{...(xiv)}$$

Equation (xiv) gives the expression for the correction to be applied. The value of tp is obtained with a thermometer. By the method of successive substitution, the value of t_{lim} is obtained.

From equations (xii) and (xiv) it is clear that in both the cases the value of

Thus, the ideal or perfect gas scale (Celsius scale) can be defined as follows:

$$\begin{matrix} t = \\ \text{(celsius} \end{matrix} \left[\frac{\begin{matrix}(PV)_t - (PV)_0 \\ \lim P \rightarrow o \lim P \rightarrow 0\end{matrix}}{\begin{matrix}(PV)_{100} - (PV)_0 \\ \lim P \rightarrow o \lim P \rightarrow 0\end{matrix}}\right] \times 100 \qquad \text{...(xv)}$$

These corrections obtained from equations (xii) and (xiv) are applicable in the range of –83°C to 400°C. Beyond 400°C. the corrections are negligibly small.

Example 7:

The pressure of air in a constant volume thermometer is 80 cm and 109.3 cm at 0°C and 100°C respectively. When the bulb is placed in hot water, the pressure is 100 cm. Find the temperature of the hot water.

Solution:

We have P_t = 100 cm of Hg

P_{100} = 109.3 cm of Hg

P_0 = 80 cm of Hg

$$t = \left(\frac{P_t - P_0}{P_{100} - P_0}\right) \times 100 = \left(\frac{100 - 80}{109.3 - 80}\right) \times 100$$

t = 68.25°C.

Example 8:

A constant volume hydrogen thermometer is used to measure the temperature of a furnace. The excess pressure in the bulb over the atmospheric pressure is found to be equal to 152 cm of Hg. At 0°C the pressure in the bulb is equal to that of the atmosphere. Calculate the temperature of the furnace, assuming that the atmospheric pressure throughout the experiment remains constant.

Solution:

We have

Here $P_0 = 76$ cm of Hg

$P_t = 152 + 76 = 228$ cm of Hg

$T_0 = 273$ K

$T = ?$

$$\frac{P_t}{T} = \frac{P_0}{T_0}$$

$$T = \frac{P_t}{P_0} \times T_0$$

$$T = \frac{228 \times 273}{76}$$

$T = 819$ K $= 546$°C.

1.15 PLATINUM RESISTANCE THERMOMETER

It is based on the principle of change of resistance with change in temperature. It was first designed by Siemen in 1871 and later on improved by Callendar and Griffiths.

A platinum resistance thermometer consists of a pure platinum wire wound in a double spiral to avoid inductive effects. The wire is wound on a mica plate. The two ends of the platinum wire are connected to thick copper leads (for lower temperatures) and connected to the binding terminals B_1 B_2. For higher temperatures the leads are of platinum, C_1 and C_2 are the compensating leads exactly similar and of the same resistance as the leads used with the platinum wire. The platinum wire and the compensating leads are enclosed in a glazed porcelain tube. The tube is sealed and binding terminals are provided at the top. The leads pass through mica discs which offer the best insulation and also prevent convection currents (Fig. 1.6).

The resistance of a wire at t°C = R_t and at 0°C = R_0. These resistances are connected by the relation

$$R_t = R_0 (1 + \alpha t + \beta t^2) \quad ...(i)$$

Here α and β are constants. The values of a and β depend on the nature of the material used. To find the values of α and β, the resistance of the platinum wire is determined at three fixed points (i) melting point of ice, (ii) boiling point of water, (iii) boiling point of sulphur 444.6°C for high temperature measurement and (iv) boiling point of oxygen – 182.5°C for low temperature measurement.

Using these values of resistance in equation (i)

$$R_{100} = R_0 [1 + \alpha 100 + \beta 100^2] \quad ...(ii)$$

and

$$R_{444.6} = R_0 [1 + \alpha 444.6 + \beta (444.6)^2] \quad ...(iii)$$

The values of α and β can be determined by solving the simultaneous equations (ii) and (iii).

From eqn. (i) $R_t = R_0[1 + \alpha t + \beta t^2]$

Neglecting βt^2 (because β is very small)

$$R_t = R_0 [1 + \alpha t] \quad ...(iv)$$

and

$$R_{100} = R_0 [1 + \alpha \times 100]$$

$$\therefore \quad R_t - R_0 = R_0 \alpha t \quad ...(v)$$

$$R_{100} - R_0 = R_0 \alpha . 100 \quad ...(vi)$$

Dividing (v) and (vi)

$$\frac{R_t - R_0}{R_{100} - R_0} = \frac{t}{100}$$

$$t = \left(\frac{R_t - R_0}{R_{100} - R_0}\right) \times 100 \quad ...(vii)$$

Knowing the values of R_0, R_{100} and R_t, t can be calculated.

The resistance of the platinum wire is found accurately using Callendar and Griffiths' bridge.

Callendar and Griffith's Bridge

This is a modified form of Wheatstone's Bridge, used to measure the change in resistance with temperature of the platinum wire used in a platinum resistance thermometer.

P and Q are two resistances of equal value and 8 is a standard fractional resistance box. The platinum wire whose resistance is, to be calculated is connected in one of the arms of the Wheatstone's bridge as shown in Fig. 1.6. The compensating leads are connected in series with the resistance S. XY is the bridge wire of length $2l$ and of resistance per unit length ρ. Let r be the resistance of the compensating leads and O the mid-point of the bridge wire. R is the resistance of the platinum wire at the temperature of the bath. Let D be the balance point, with some suitable value of S, x cm to the left of O.

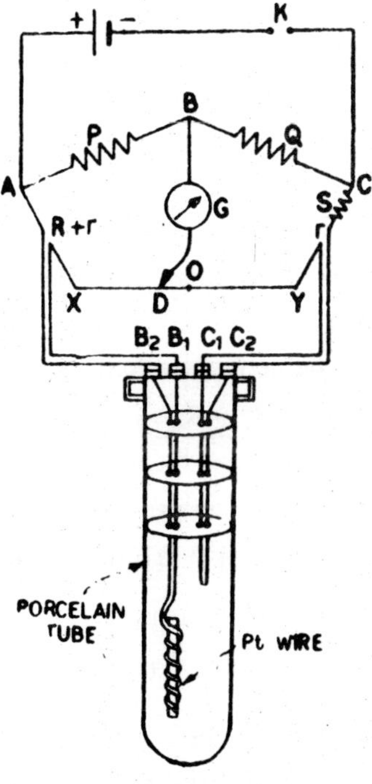

Fig. 1.6

Then length $XD = l - x$

$DY = l + x$

As P = Q, the resistance of the arm AD = resistance of the arm CD

$$\therefore \quad R + r + (l - x)\rho = S + r + (l + x)\rho$$

$$R + r + l\rho - x\rho = S + r + l\rho + x\rho$$

$$R = S + 2x\rho.$$

(If the balance point is obtained to the right of O, it can be shown that R = S + 2xρ).

Thus, using a Callendar and Griffith's Bridge the resistance R_t of the platinum wire can be determined at temperature t,

A scale can be fixed along the bridge wire XY to read platinum scale temperatures directly. The range will be different for different values of S.

Correction : The temperature t measured on the platinum scale according to equation (vii) is not accurate because βt^2 term has been neglected. Therefore, the correct thermodynamic temperature θ will differ from t.

Here $$\theta - t = \delta\left[\left(\frac{\theta}{100}\right)^2 - \left(\frac{\theta}{100}\right)\right] \qquad \text{...(viii)}$$

The value of δ can be calculated by knowing the values of α and β.

Equation (viii) can be derived as follows :

$$\theta - t = \theta - \left(\frac{R_t - R_0}{R_{100} - R_0}\right) \times 100$$

Take $$Rt = R_0(1 + \alpha\theta + + \beta\theta^2)$$

and $$R_{100} = R_0[1 + \alpha 100 + \beta(100)^2]$$

$$\therefore \quad \frac{R_t - R_0}{R_{100} - R_0} = \frac{\alpha\theta + \beta\theta^2}{\alpha 100 + \beta(100)^2}$$

$$\therefore \quad \theta - t = \theta - \left(\frac{\alpha\theta + \beta\theta^2}{\alpha + \beta 100}\right)$$

$$\theta - t = \frac{\alpha\theta + 100\beta\theta - \alpha\theta - \beta\theta^2}{\alpha + \beta 100}$$

$$\theta - t = \frac{100\beta\theta - \beta\theta^2}{(\alpha + \beta 100)}$$

$$\theta - t = \frac{-(100)^2\beta}{(\alpha + \beta 100)}\left[\frac{\theta^2}{(100)^2} - \frac{\theta}{100}\right]$$

$$\theta - t = \delta\left[\left(\frac{\theta}{100}\right)^2 - \left(\frac{\theta}{100}\right)\right] \quad ...(ix)$$

Here $$\delta = \frac{-(100)^2\beta}{(\alpha + \beta 100)}$$

The value of δ for pure platinum is approximately equal to 1.5.

Advantages: The platinum resistance thermometer can be used with great accuracy to measure temperatures ranging from –200°C to –1200°C. Its accuracy is 0.1°C. Once the platinum resistance thermometer has been standardised with a constant volume hydrogen thermometer, it can be used as a standard thermometer.

Its main disadvantage is that the adjustment of the Callendar and Griffith's' bridge takes a long time. Moreover, the wire may not attain the temperature of the bath in a short time.

Example 9:

The resistance of the platinum wire of a platinum resistance thermometer at the ice point is 5 ohms and at the steam point 5.93 ohms. The pressure exerted by the gas in a constant volume gas thermometer is (i) 100 cm of Hg at ice point (ii) 136.6 cm of Hg at the steam point. When both the thermometers are inserted in a hot bath the resistance of the platinum wire is 5.796 ohms and the pressure of the gas is 131.11 cm of Hg. Calculate Celsius temperature of the bath (i) on the platinum scale and (ii) on the gas scaled

Solution:

We have

(i) On the platinum scale

R_0 = 5 ohms

R_{100} = 5.23 ohms

R_t = 5.795 ohms

$$t\rho = \left(\frac{R_t - R_0}{R_{100} - R_0}\right) \times 100$$

$$= \left(\frac{5.795 - 5}{5.93 - 5}\right) \times 100 = 85.48°C$$

(ii) On the gas scale

$$P_0 = 100 \text{ cm Hg}$$

$$P_{100} = 136.6 \text{ cm Hg}$$

$$P_t = 131.11 \text{ cm Hg}$$

$$t = \left(\frac{P_t - P_0}{P_{100} - P_0}\right) \times 100$$

$$= \left(\frac{131.11 - 100}{136.6 - 100}\right) \times 100 = 85°C$$

Example 10:

If the platinum temperature, corresponding to 50°C on the gas scale is 50.25°, what will be the temperature on the platinum scale corresponding to 150°C on the gas scale.

Let θ be the temperature on the gas scale and t on the platinum scale.

Let $$\theta - t = \delta\left[\left(\frac{\theta}{100}\right)^2 - \left(\frac{\theta}{100}\right)\right]$$

$$\theta = 50 \; ; \; t = 50.25$$

$$\therefore 50 - 50.25 = \delta\left[\left(\frac{50}{100}\right)^2 - \left(\frac{50}{100}\right)\right] \quad \text{...(i)}$$

(ii) $\theta = 150°; \; t = x$

$$\therefore \; 150 - x = \delta\left[\left(\frac{150}{100}\right)^2 - \left(\frac{150}{100}\right)\right] \quad \text{...(ii)}$$

Dividing (ii) by (i) and simplifying

$$x = 149.25°C$$

Example 11:

The resistance of a platinum wire at 0°C, 100°C and 444.6°C is found to be 5.5, 7.5 and 14.5 ohms respectively. The resistance of a wire at a temperature t°C is given by the equation

Solution:

We have $R_t = R_0 (1 + \alpha t + \beta t^2$

Find the values of α and β.

Here $R_0 = 5.5$ ohms

$R_{100} = 7.5$ ohms

$R_{444.6} = 14.5$ ohms

$R_t = R_0 (1 + \alpha t + \beta t^2)$

$\therefore \quad R_{100} = R_0 [1 + \alpha(100) + \beta(100)^2)$ ...(i)

$R_{444.6} = R_0 [(1 + \alpha(444.60 + \beta(444.6)^2]$ ...(ii)

∴ Substituting the values in equations (i) and (ii)

$7.5 = 5.5 [1 + 100\alpha + (100)^2 + \beta]$ ...(iii)

$14.5 = 5.5 [1 + 444.6\alpha + (444.6)^2 + \beta]$ ...(iv)

Simplifying equations (iii) and (iv)

$$\alpha + 100\beta = \frac{2}{5.5 \times 100} \quad \text{...(v)}$$

$$\alpha + 444.6\,\beta = \frac{9}{5.5 \times 444.6} \quad \text{...(vi)}$$

Subtracting equation (v) from (vi)

$$344.6\,\beta = \frac{900 - 2 \times 444.6}{5.5 \times 444.6 \times 100}$$

$$\therefore \quad \beta = 1.281 \times 10^{-7}/°C$$

Substituting the value of β in equation (v)

$$\alpha + 100\,(1.281 \times 10^{-7} = \frac{2}{5.50 \times 100}$$

$$\alpha = 363.6 \times 10^{-5} - 1.281 \times 10^{-5}$$

$$\alpha = 362.319 \times 10^{-5}$$

$$\alpha = 3.62 \times 10^{-3}/°C$$

1.16 SEEBECK EFFECT

In 1821, Seebeck found that a current flows in a circuit consisting of two dissimilar metals when one junction is heated while the other junction is kept cold. This was a remarkable experiment because *no cell* was used in the circuit. He connected a plate of bismuth between copper wires connected to a galvanometer [Fig. 1.7(i)]. He found that if one of the junctions was heated while the other was kept cold, then a current flowed through the galvanometer. He repeated his experiment by taking a thermocouple of Fe and Cu [Fig. 1.7 (ii)].

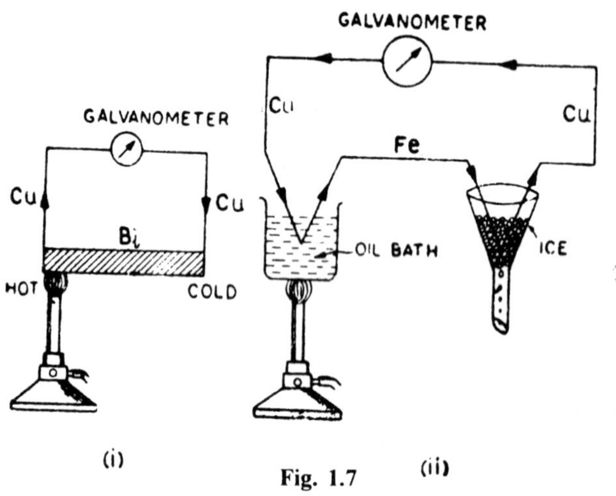

Fig. 1.7

If both the junctions are at 0°C, there is no deflection in the galvanometer. When one of the junctions is kept at 0°C, *i.e.*, at the temperature of melting point of ice and the other junction is heated gradually, current flows in the circuit. It was found that current flows from copper to iron at the hot junction and iron to copper at the cold junction. The current increases until the hot junction is at a temperature of 270°C. If the heating is continued beyond 270°C, the current decreases and finally at 540°C the current is zero.

It was discovered by *Cumming* in 1823 that, on increasing the temperature of the hot junction beyond 540°C, the direction of the current is reversed. It flows from iron to copper through the hot junction and copper to iron through the cold junction.

The current produced in this way without the use of a cell or a battery is known as *thermoelectric* current and this branch of electricity is known as thermoelectricity. The effect is known as *Seebeck effect.*

The temperature of the hot junction at which maximum current flows in a circuit is known as *neutral temperature* for that couple. The neutral temperature for a given thermocouple is fixed and remains constant whatever may be the temperature of the cold junction. The E.M.F. produced in this way is called *themo-E.M.F.*

If a graph is plotted between the temperature of the hot junction and the thermo-E.M.F., the cold junction being kept at 0°C the graph is a

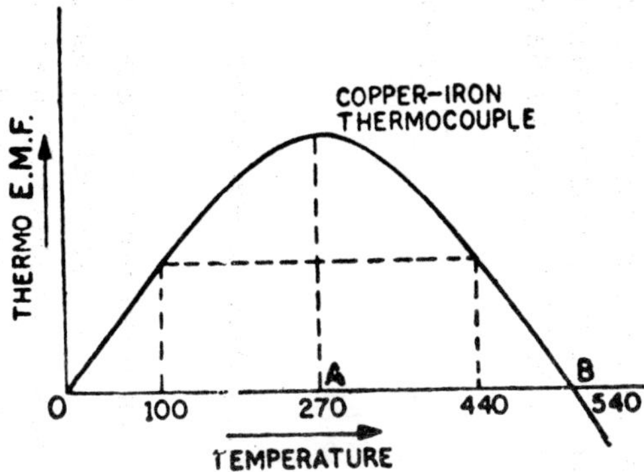

Fig. 1.8

parabolic curve. The thermo-E.M.F. E varies with temperature according to $E = at + bt^2$, where a and b are constants. The point A represents the neutral temperature. The point B is the *temperature of inversion*, beyond which the current is reversed (Fig. 1.8).

The temperature of inversion is not fixed. It is as much above the neutral temperature as the cold junction is below the neutral temperature.

Suppose the temperature of the cold junction = θ_1

Neutral temperature = θ_n

Temperature of inversion = θ_2

Then, $\theta_n - \theta_1 = \theta_2 - \theta_n$, or $\theta_n = \left(\dfrac{\theta_1 + \theta_2}{2}\right)$

In the case of Cu–Fe thermocouple, if the cold junction is at 100°C temperature of inversion = 440°C.

1.17 THERMO-ELECTRIC THERMOMETER

It is based on the principle of Seebeck effect. Two dissimilar metals are selected. The thermo-E.M.F. E reduced, depends upon the difference of temperature between the of and the cold junctions. For measuring temperatures up to 300°C, copper-constantan thermocouple is used. Iron-nickel is used between 300°C and 600°C and nickel-nichrome is used between 600°C and 1000°C. For temperatures between 1000°C and 1600°C, a thermo-couple of platinum and an alloy of platinum and rhodium is used. Beyond 1600°C, up to 2000°C iridium and an alloy of iridium and rubidium is used. For measuring temperatures between 2000°C and 3000°C, tungsten and molybdenum thermocouple is used.

The hot junction is made by welding the junction of two metals and the junction is enclosed in a fine capillary tube. The whole arrangement is enclosed in an outer porcelain tube (Fig. 1.9). The wires 1 and 2 inside the tube are kept separate with the help of mica strips.

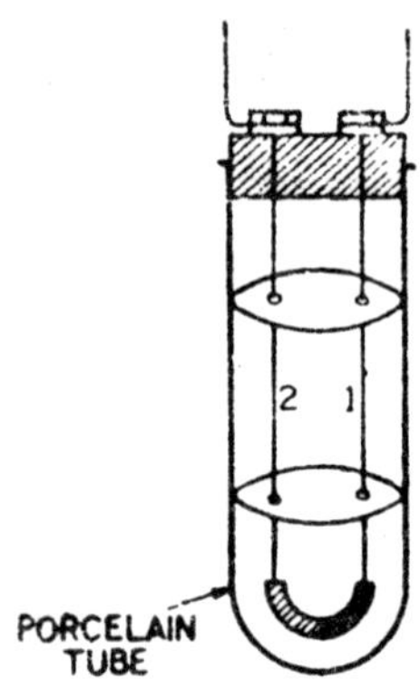

Fig. 1.9

Measurement of Temperature

The most important part of the experiment for the measurement of temperature is the accuracy with which the thermo-E.M.F. can be measured. Two methods are usually employed, *e.g.*

(i) Using a sensitive galvanometer,

(ii) Using potentiometer.

(i) **Using a Sensitive Galvanometer :** A sensitive galvanometer is connected with the thermocouple (Fig. 1.10). One of the junctions is kept in melting ice (0°C) and the other junction in an oil bath. The oil bath is heated and thermo-E.M.F. is produced in the circuit. The temperature of the oil bath is determined by using a constant volume hydrogen thermometer or some other sensitive thermometer.

Corresponding to the various temperatures of the hot junction (oil bath) the corresponding deflections in the galvanometer are observed. A graph is plotted between the deflection along the y-axis arid the temperature of the hot junction along the x-axis (Fig. 1.11). Thus, for the given thermocouple the galvanometer is calibrated.

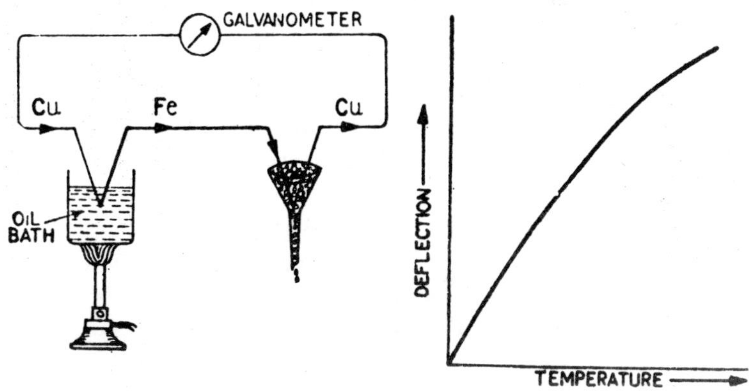

Fig. 1.10 Fig. 1.11

The bath whose temperature is to be measured, is taken. The hot junction of the thermocouple is immersed in the bath, other junction being kept in melting ice. The deflection produced in the galvanometer is observed. From the graph, the corresponding value of temperature is determined. Moreover, the scale of the galvanometer can be directly marked in °C to show the temperatures directly. Such a thermoelectric thermometer is not very sensitive.

(ii) **Using a Potentiometer:** The thermo-E.M.F. produced in a thermocouple is of the order of 10^{-5} volt. Therefore the

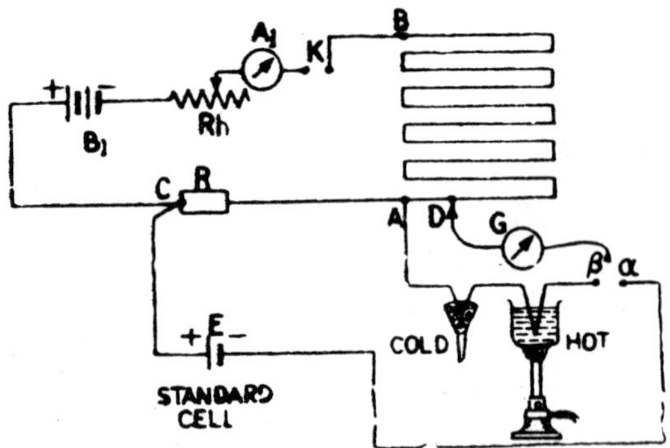

Fig. 1.12

potentiometer should be sensitive to measure up to 10^{-6} volt. For this purpose the circuit used is as shown in Fig. 1.12. To check the correctness of the connections, first close the key α. If on closing the key K, the deflection in the galvanometer decreases, then the connections are correct.

The standard cadmium cell E has an E.M.F. = 1.0183 volts. The potentiometer has ten wires of length one metre each. The potentiometer whose wire has a resistance of one ohm per metre is used. The distance AD = 30 cm. The jockey is kept at D.

A resistance of 1018 ohms is introduced in B and the key a is closed. The rheostat Rh is adjusted so that the deflection in the galvanometer is zero. Then, the P.D. across CD = 1.0183 volts.

The resistance of AD $= \dfrac{1 \times 30}{100} = 0.3$ ohm.

The resistance of CD = 1018 + 0.3 1018.3 ohm.

Fall of potential for one ohm-resistance

$$= \frac{1.0183}{1018.3} = 10^{-3} \text{ volt.}$$

The potentiometer wire has a resistance of one ohm per metre. Therefore the fall of potential for one metre length of the potentiometer wire = 10^{-3} volt and the fall of potential per mm length of the potentiometer wire

$$= \frac{10^{-3}}{1000} = 10^{-6} \text{ volt per mm.}$$

Thus, the potentiometer wire is calibrated.

The key α is taken out and the key β is introduced. For the various temperatures of the hot junction, the corresponding balance points are observed. The lengths are measured from the point A. Thus the corresponding thermo- E.M.F. produced in the circuit can be calculated. The temperatures of the hot junction are measured with a standard constant volume hydrogen thermometer. A graph is plotted between the thermo-E.M.F. produced and the corresponding temperature. The potentiometer wire can be directly calibrated to measure the temperatures.

The hot junction is kept in the bath of the furnace whose temperature is to be measured and the balance point is found. From the balancing length, the corresponding temperature is either read from the graph or from the chart provided.

For practical purposes charts are provided with each thermocouple.

Advantages

1. It has a wide range and can measure temperatures from –200°C up to 3000°C.
2. It is useful to measure rapidly varying temperatures.
3. It can be easily constructed and it is cheap.
4. It is useful to measure the temperature of hot furnaces and does not require any calculations, once it has been calibrated.

Limits and Drawbacks:

This thermometer does not give accurate temperatures over a wide range. Different thermocouples are to be used for different ranges. The potentiometer is to be calibrated for different thermocouples separately. Moreover, the neutral temperature of a thermocouple also limits its range.

1.18 HELIUM VAPOUR PRESSURE THERMOMETER

This thermometer is used to measure temperature up to 0.7K. The apparatus consists of a bulb A containing liquid helium. This bulb is connected to the manometer limbs M_1 and M_2, through a connecting tube C. The tube C is surrounded by a copper tube B to ensure uniform temperature of the vapour, it is a reservoir containing mercury (Fig. 1.13).

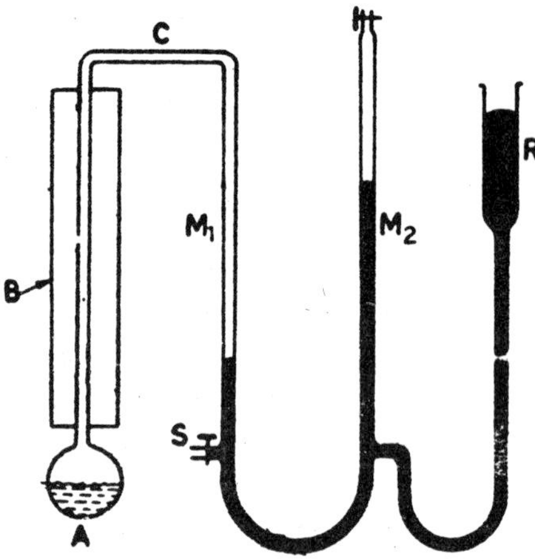

Fig. 1.13

Initially the reservoir B is lowered so that the mercury in the manometers M_1 and M_2 is below the stop-cock S. The tube is connected to an evacuation pump to remove air in the tube C and the bulb A. The stop-cock S is closed after evacuation and the bulb A is placed in the bath whose temperature is to be measured. The pressure of saturated

Table 1.1 : Vapour pressure of helium (He^4) at different temperatures.

Temp. K	Pressure mm of Hg
5.00	1460
4.50	980
4.00	016
3.50	353
3.00	181
2.50	77
2.00	23.2
1.60	3.6
1.00	0.12
0.40	1.6×11^{-5}
0.10	3.4×10^{-32}

helium vapour is measured from the difference in levels of Hg in the limbs M_1 and M_2. With the help of constant tables, giving the vapour pressure of the liquid at various temperatures, the temperature corresponding to any observed vapour pressure is determined. For lower temperatures, the graph between saturated vapour pressure and temperature is extrapolated.

1.19 STANDARDIZATION AND TEMPERATURE SCALE

The various thermometers depend upon the property of a substance. The temperature scale shown by a thermometer will also depend on the property of the substance selected. Due to this reason, the scale of temperature is arbitrary depending on the property of the substance. The thermometers thus manufactured agree only at the fixed points and not at any other temperature.

Let the magnitude of the property selected be

x_t, x_{ice}, x_{steam} at t°C, 0°C and 100°C

The temperature t can be calculated from the relation

$$\frac{t}{100} = \frac{x_t - x_{ice}}{x_{steam} - x_{ice}} \quad \text{...(i)}$$

It means that the temperature scale selected and the property of the substance chosen vary uniformly. But it has been found that the properties of a substance *e.g.*, expansion with rise in temperature in mercury thermometer, change of resistance with temperature in a platinum resistance thermometer etc. do not vary uniformly at all temperatures. Due to this reason, all the thermometers do not show the same temperature readings at one fixed temperature. Suppose the temperature of the liquid is measured by:

(i) a mercury thermometer

(ii) a constant volume hydrogen thermometer and

(iii) a platinum resistance thermometer at the same time. All these thermometers will not show the same reading. The readings may be 40.11°C, 40°C, and 40.36°C. Thus it is found that the temperature shown by the thermometer depends, upon the nature of the thermometer used. However, these thermometers will show the same temperatures at fixed points *e.g.*, melting point of ice and boiling point of water.

Comparison of different thermometers.

Constant volume hydrogen thermometer	Pt. resistance thermometer	Mercury thermometer
0°C	0°C	0°C
20°C	20.24°C	20.09°C
60°C	60.36°C	60.09°C
100°C	100°C	100°C

In 1887, an international committee of scientists suggested that a scale of temperature shown by constant volume hydrogen thermometer be taken as a standard scale, 0°C being that of melting ice and 100°C being the temperature of steam at normal pressure. But hydrogen also is not a standard or a perfect gas.

Later on it was thought that temperature scale suggested by Kelvin be used as a standard scale because it does not depend upon property of a substance. The Kelvin scale of temperature (K) (work scale) agrees with the absolute gas scale.

As it appears difficult to realize the Kelvin scale in practice it is suggested in 1933 that an international scale of temperature be adopted. The melting point of ice and boiling point of pure water at normal pressure etc. are given the fixed values and the temperatures between these points are calculated by a specific formula. The international scale of temperature is the nearest practical approach to the Kelvin scale of temperature.

Fixed Points

1. **Oxygen Point :** Temperature of equilibrium between liquid oxygen and gaseous oxygen at normal pressure = –182.97°C.
2. **Ice Point :** Temperature of equilibrium between ice and air saturated water at normal pressure = 0.000°C.
3. **Steam Point :** Temperature of equilibrium between liquid water and its vapour at normal pressure = 100.000°C.
4. **Sulphur Point :** Temperature of equilibrium between liquid sulphur and its vapour at normal pressure = 444.60°C.
5. **Silver Point :** Temperature of equilibrium between solid silver and liquid silver at normal pressure = 960.8°C..

6. **Gold Point :** Temperature of equilibrium between solid gold and liquid gold at normal pressure = 1063°C.

The scale is divided into four parts for the purpose of determining the temperature between two fixed points:

(a) **From –190°C to 0°C :** The platinum resistance thermometer is used and the relation is

$$R_t = R_0\,[1 + \alpha t + \beta t^2 - \gamma\,(t - 100)\,t^2].$$

(b) **From –0°C to 660°C :** The platinum resistance thermometer is used and the relation is

$$R_t = R_0\,(1 + \alpha t + \beta t^2).$$

(c) **From –660°C to 1063°C :** The thermocouple of platinum and an alloy of platinum-rhodium is used. One junction is kept in melting ice. The thermo-E.M.F. produced is measured and the relation used is

$$E = \alpha + \beta t + \gamma t^2.$$

(d) **Beyond 1063°C :** The ratio of the intensities of monochromatic heat radiation emitted by the black body at t°C and at 1063°C is determined and the temperature is calculated using Planck's radiation law.

1.20 ABSOLUTE ZERO AND ICE POINT

For approximate purposes, the value of absolute zero or zero degree Kelvin is taken as –273°C. But near about absolute zero its correct determination is a necessity. For accurate work the value of absolute zero is taken as –273.16°C. Similarly, the temperature of melting point of ice was suggested by the International Advisory Committee of Thermometry in 1948, as 273.15 K. But in 1954, the international committee fixed 273.16 K (triple point of water) as ice point. Thus for accurate purposes.

Ice point	=	0°C = 273.16 K
Steam point	=	100°C = 373.16 K.
Zero degree Kelvin	=	–273.16°C.

1.21 LOW TEMPERATURE MEASUREMENT

The various thermometers used to measure low temperatures are discussed below:

1. **Liquid Thermometers :** With alcohol thermometer temperatures up to –100°C can be measured. With mercury in glass, temperatures up to –30°C can be measured. The accuracy is about 0.1°C. These are convenient and their response a quick.
2. **Gas Thermometers :** Constant volume hydrogen thermometer can be used to measure up to –250°C and with constant volume helium thermometer temperatures up to –268°C can be determined. These thermometers are quite accurate but they are quite bulky.
3. **Resistance Thermometers:** With platinum resistance thermometers temperatures up to –190°C can be measured accurately. Its accuracy is 0.01°C. It is quite accurate but its response is slow.
4. **Thermoelectric Thermometers:** With copper-constantan thermocouple or platinum and silver thermocouple, temperatures up to –250°C can be measured. Its accuracy is 0.05°C.
5. **Vapour Pressure Thermometers :** These can be used to measure up to –268°C. Below –268°C, helium vapour pressure thermometer is used. It can be used up to –272°C.
6. **Magnetic Thermometer:** Near about the absolute zero temperature, magnetic thermometers are used. They are based upon the principle of change in susceptibility with temperature according to Curie's Law.

1.22 HIGH TEMPERATURE MEASUREMENT

The various thermometers used to measure high temperatures are given below:

1. **Liquid Thermometers :** Mercury in glass can measure up to 300°C. If the space above mercury is filled with some inert gas like nitrogen or helium, it can be used up to 600°C.
2. **Gas Thermometers :** Constant Volume hydrogen thermometer having a platinum bulb, can be used to measure up to 500°C. With a porcelain bulb temperatures up to 1100°C can be measured. Beyond 1100°C, using nitrogen in place of hydrogen, temperatures up to 1500°C can be measured. They are quite bulky for use.
3. **Resistance Thermometers :** The platinum thermometer can be used up to 1200°C when properly calibrated. Its accuracy is 0.1°C. It is a slow measuring instrument.

4. **Thermoelectric Thermometers :** For measuring up to 300°C copper-constantan thermocouple is used. Iron-nickel thermocouple is used between 300°C and 600°C. Nickel-nichrome thermocouple is used between 600°C and 1000°C. A thermocouple of platinum and rhodium is used between 1000°C and 1600°C. Between 1600°C and 2000°C a thermocouple of iridium and an alloy of iridium and rubidium is used. For measuring between 2000° and 3000°C, tungsten and molybdenum thermocouple is used.

5. **Pyrometers :** For measuring the temperature of furnaces and the sun, optical pyrometers are used.

EXERCISES

1. Write short notes on:
   (i) Callendar's compensated constant pressure air thermometer.
   (ii) Platinum resistance thermometer.
   (iii) Constant volume hydrogen thermometer.
   (iv) Thermoelectric thermometer.
   (v) Measurement of high temperature.
   (vi) Measurement of low temperature.
   (vii) Rankine and Kelvin scales of temperature.
   (viii) Errors and corrections in mercury thermometers.
   (ix) Advantages of gas thermometers.
   (x) Universal gas constant.
2. The temperature of a furnace is 2000°C. What is this temperature:
   (i) on the Rankine scale and
   (ii) on the Kelvin scale?
3. The normal boiling point of liquid hydrogen is –253°C. What is this temperature:
   (i) on the Kelvin scale and
   (ii) on the Rankine scale?
4. At what temperature will the Kelvin scale reading be double the Fahrenheit reading?
5. Find the value of the universal gas constant.

6. Find the value of the ordinary gas constant for nitrogen.
7. The resistance of a platinum resistance thermometer at the ice point is 5 ohms and at the steam point is 6.93 ohms. The pressure exerted by the gas in constant volume gas thermometer at the ice point is 100 cm of Hg and at the steam point it is 1366cm of Hg. When both the thermometers are placed in a bath, the resistance of a resistance thermometer is 5.795 ohms and the pressure of the gas is 114.9 cm of Hg. Calculate the Celsius temperature of the liquid:

   (i) on the platinum scale, and

   (ii) on the gas scale.
8. The bulb of the Callendar's compensated constant pressure air thermometer is 800 cm^3. When the bulb is immersed in a bath, 200 cm^3 of mercury has to be drawn out of the reservoir. Calculate the temperature of the bath on the Celsius scale.
9. If the platinum temperature corresponding to 60°C on the gas scale is 60.25°C, what will be the temperature on the platinum scale corresponding to 120°C on the gas scale?
10. If the platinum temperature corresponding to 60°C on the gas scale is 60.36°C, what is the platinum scale temperature corresponding to 151.7°C on the gas scale?
11. Write an essay on the measurement of high and low temperatures.
12. State with reasons the type of thermometer which you consider most suitable for use at temperatures (a) –250°C (6) 700°C and (c) 2000°C. Indicate briefly the methods of their use.
13. Describe a platinum resistance thermometer. Explain how it works with the help of Callendar and Griffith's bridge. How does the platinum temperature of a body differ from its true temperature?
14. What do you understand by the absolute scale of temperature? Is the negative temperature possible on this scale ?
15. Describe the Callendar and Griffith's method of determining the temperature coefficient of platinum. In what respects it is superior to Carey Foster's method? What do you understand by:

   (i) platinum leads,

   (ii) compensating arm,

(iii) non-inductive winding of the platinum wire?

16. Give the theory and construction of a constant volume gas thermometer. In what respects is this thermometer superior to mercury in glass thermometer.
17. Give an account of the thermoelectric thermometry and discuss the range, sensitivity and usefulness of some important thermocouples.
18. Describe carefully the methods for measuring low temperatures in the range –100°C to –273°C. Explain the concept of absolute zero of temperature.
19. State the advantages of using a permanent gas as a thermometric substance. Describe the working of a constant volume hydrogen thermometer.
20. Describe a resistance thermometer. Explain how it is used to measure temperatures accurately. Discuss its advantages over a thermo-elcctric thermometer.
21. Describe and explain how the platinum resistance thermometer can be used to measure temperatures accurately in the region 200°C to 500°C.
22. Give the construction of a thermoelectric thermometer and compare its performance with a standard gas thermometer. What precautions should be taken in measuring temperatures with a thermoelectric thermometer?
23. Describe the construction and working of a standard gas thermometer.
24. What is meant by a scale of temperature? On what does the definition of any particular scale depend?
25. Give a detailed account of the experiment you would perform to determine the temperature of boiling aniline on the platinum resistance scale of temperature. By how much does the temperature measured by this thermometer differ from that measured by gas thermometer?
26. Discuss the advantages of using one of'the permanent gases as a thermometric substance for denning a scale of temperature. Describe some convenient and accurate form of gas thermometer. Explain its mode of use and show bow the temperature is calculated from the observations made with it.

27. State briefly the principles underlying the working of the following thermometer:
    (i) Constant volume gas thermometer.
    (ii) Resistance thermometer.
    (iii) Thermo-electric thermometer
    (iv) Vapour-pressure thermometer.
28. Describe Calleodar and Griffith's bridge for measuring the resistance of a platinum resistance thermometer at various temperatures. Deduce by how much a temperature measured by this thermometer differs from that measured by a gas thermometer.
29. Describe a platinum resistance thermometer. How would you calibrate and use it for measuring the temperature of a body? Mention the advantages of this thermometer.
30. On what principle a the working of a platinum resistance thermometer based? Describe Callendar and Griffiths bridge for the accurate measurement of resistance. How is the true temperature deduced from the measured platinum temperature?

2

Thermal Expansion

2.1 GENERAL

We know from our everyday experience that material bodies (whether solids or fluids) undergo, when heated, a change in dimensions. This change, however small, is an increase or expansion in length, area or volume. There are just a few notable exceptions: like re-solidified iodide of silver (below 142°C), some Ne-Fe alloys and water (also some aqueous solutions) whose expansion is anomalous.

Thermal expansion of fluids is very much greater. Liquids expand more than the solids and gases expand even more than the, liquids. Solids and liquids are regarded as incompressible, their volume remains practically unaffected by pressure while a gas is quite easily compressed hence its volume depends on pressure also. For the sake of convenience and clarity, we shall deal with the expansion of solids, liquids and gases under separate sections.

SECTION I: EXPANSION OF SOLIDS

Solids may be divided into two distinct categories:

(a) isotropic, and

(b) anisotropic.

(a) *Isotropic Solids* are those whose physical properties (like refractive index, thermal expansion, thermal conductivity, electrical resistance, etc.) are identical in all directions so that when heated, they expand proportionally in every direction. Thus, while they undergo a change in area or volume, they continue to preserve their shape.

All amorphous solids (solids with no determinate shape or external crystalline form) with a regular pattern and metals (composed of large number of minute crystals thus have their average property the same in every direction due to a random orientation of these constituent crystals) come under this class.

(b) *Non-isotropic or Anisotropic Solids* are those solids whose physical properties are not same in all directions. So that when heated their expansion is not only different in different directions but may also be of different signs (*i.e.*, an expansion in one direction and a contraction perpendicular to it), hence they undergo a change in shape. All crystalline solids and substances like wood belong to this class.

Expansion of Isotropic Solids

Since a solid has length, breadth and thickness therefore it has a surface (area) and a volume also. There are thus three types of expansion (i) linear expansion, (ii) superficial expansion and (iii) cubical expansion.

2.2 LINEAR EXPANSION

Since an isotropic solid expands equally in all directions we take one dimension, length, for considering linear expansion. Linear expansion is defined as the fractional change in length produced per degree rise of temperature. Consider a bar of original length l_0 at $l_0°$ and has a length l_t at t° then l_t at l_0 will be the increase in length for $(t - t)_0°$ rise in temperature (conventionally in° C or K).

This increase in length, though depends on the nature of the material of the bar, is found to be proportional to (i) the length of the bar and (ii) to the rise in its temperature.

i.e., $(l_t - l_0) \propto l_0 (t - t_0)$

or $(l_t - l_0) = l_0 \alpha (t - t_0)$

where a is a constant which depends on the nature of the material (not on shape or size) is called the coefficient of linear expansion (C.L.E.).

The expression $\alpha = \dfrac{l_t - l_0}{l_0\,(t - l_0)}$ is then *for the mean coefficient of linear expansion* in the temperature range between t_0, and t. If the difference $t - t_0$ is not too large we can take, for $t > t_1$ and $t_2 > t_0$

$$l_1 = l_0\,[1 + \alpha\,(t_1 - t_0)]$$

$$l_2 = l_0\,[1 + \alpha\,(t_2 - t_0)]$$

which on simplification yields true coefficient of expansion at t° as

$$\alpha = \frac{1}{l}\,\frac{\delta l}{\delta t}$$

where l is original length and δl is the change in the length due to rise δt in temperature (which is extremely small). We can also define the zero coefficient of linear expansion as

$$\alpha_z = \frac{1}{l_0}\,\frac{dl}{dt}$$

But the experimental observations yield mean coefficient over different temperature ranges which are not usually much different from a_z*.

α is measured in K^{-1} or $°C^{-1}$. It does not depend on the unit of length chosen, but is a function of temperature given by

$$\alpha \equiv \alpha_z + \alpha_1 t + \alpha_2 t^2 + \ldots$$

with $\alpha_z > \alpha_1 > \alpha_2 > \alpha_3$ etc. whence for all practical purposes $\alpha \cong \alpha_z$, (unless very high changes in temperature are involved or extremely high precision is aimed at).

2.3 SURFACE OR SUPERFICIAL EXPANSION

When a solid body is heated, it increases in length as well as in breadth, so that its surface area also increases. This increase in surface area is called its surface area, or superficial expansion. The coefficient of superficial expansion of a solid may by defined as fractional increase in area per degree rise in temperature. It is usually denoted by p and is measured as K^{-1} or $°C^{-1}$. It again depends on the nature of the material but is independent of the system of units used, for length/area.

To define it mathematically, if Ay be the original surface area at $t_0°$ and A_1 be the surface area at t° we have (for $t - t_0 = \delta t$)

$$\beta = \frac{A_t - A_0}{A_0(t - t_0)}$$

or $$A_t = A_0[1 + \beta(t - t_0)]$$

or, with sufficient accuracy, when $t \to t_0$

$$\beta = \frac{1}{A_0}\frac{\delta A}{\delta t}$$

2.4 CUBICAL EXPANSION

The increase in volume is known as cubical expansion. We may define the coefficient of cubical expansion of a solid as the increase in volume per unit volume per degree (K or °C) rise of temperature. Here again it is independent of the units used for measuring volume (but of course depends on temperature unit which we invariably take Celsius ≡ Kelvin) and is represented by y, K^{-1} or $°C^{-1}$.

Thus, if V_0 be the original volume of the solid at $t_0°$ and V, be the volume at t° (t being very close to t_0), we get

$$\gamma = \frac{V_t - V_0}{V_0(t - t_0)} \lim_{\delta t \to 0} \frac{1}{V_0}\frac{\delta V}{\delta t} \quad (\text{where } \delta t = t - t_0)$$

or, $$V_t = V_0[1 + \gamma(t - t_0)]$$

It may be noted here that the expansion of a hollow vessel or a tube is the same as that of a solid of the same material, size and shape.

2.5 RELATION BETWEEN THE THREE COEFFICIENTS OF EXPANSION

In the case of superficial expansions we have linear expansions in two dimensions while in cubical expansion we have three linear expansions in three dimensions.

Superficial Expansion, β : For simplicity consider a square plate of metal of unit length at $t_0°$ and let it be heated by 1° so that $t - t_0 = 1°$. Then the length of each side will become $(1 + \alpha)$. The new area will thus be $(1 + \alpha)^2$.

∴ $$1 + \beta = (1 + \alpha)^2$$

or $$\beta = 2\alpha + a^2 \cong 2\alpha \qquad (\because \alpha^2 \text{ is negligible})$$

i.e., coefficient of superficial expansion of the material is approximately equal to twice its coefficient of linear expansion.

Cubical Expansion, γ: Consider a cube of metal, the length of each edge being unity. Let its temperature be raised by 1° then each side of 'the cube will become (1 + α). So, volume of the cube will be, from definition

$$(1 + \alpha)^3 = 1 + \gamma$$

or, $$y = 3\alpha + 3\alpha^3 + \alpha^3$$

$$\cong 3\alpha \qquad (\because 3\alpha^2 \text{ and } \alpha^3 \text{ are negligible})$$

i.e., the coefficient of cubical expansion is approximately thrice as much as its linear expansion coefficient.

The two results together give

$$\gamma : \beta : \alpha : : 3 : 2 : 1$$

2.6 RELATION BETWEEN CUBICAL EXPANSION AND DENSITY

We know that mass of a substance is conserved and is quite independent of its temperature. It is given by

$$m = V \, . \, \rho \text{ or } \rho = \frac{m}{V}$$

where p is the density and V is the volume. But volume depends on the temperature by

$$V_t = V_0 \left[1 + \gamma \left(t - t_0\right)\right]$$

where Vo is the true volume of substance hence the density at t_0° (or ideally at 0° will be given by

$$\rho_0 = \frac{m}{V_0}$$

∴ density at t° = ρ_i ≡ say ρ

or $$= \frac{m}{V_t} = \frac{m}{V_0\left[1 + \gamma\left(t - t_0\right)\right]}$$

or $$\rho_t = \frac{m}{V_0} = \left[1 + \gamma\left(t - t_0\right)\right]^{-1}$$

i.e., $\rho_t \cong \rho_0 [1 - \gamma (t - t_0)]$

or, also $\rho_0 = \rho_t [1 + \gamma (t - t_0)]$

These relations hold good for both solids and fluids and enable us to determine the value of γ tor a substance from the knowledge of its densities at two different temperatures, (if, does not change within this range so temperature). Thus, if V_1, V_2 and ρ_1, ρ_2 be volumes and densities of substance of mass m at temperatures t_1 and t_2° C, we have

$$m = V_1\rho_1 = V_1\rho_2$$

or
$$\frac{\rho_1}{\rho_2} = \frac{V_2}{V_1} = \frac{V_0(1 + \gamma t_2)}{V_0(1 + \gamma t_1)}$$

or
$$\frac{\rho_1}{\rho_2} = 1 + \gamma t_2 - \gamma t_1 - \gamma^2 t_1 t_2$$

$$\cong 1 + \gamma (t_2 - t_1)$$

$$\therefore \quad \gamma = \frac{\rho_1 - \rho_2}{\rho_2 (t_2 - t_1)}$$

2.7 EXPERIMENTAL DETERMINATION OF THE COEFFICIENT OF LINEAR EXPANSION OF A SOLID

There are essentially two methods for measuring this small quantity for a solid, *viz.*: (i) end measurement, by measuring the whole length of its rod from one end to the other, and (ii) line measurement, by measuring the distance between two lines or marks engraved on the rod, before and after it has been heated through a known temperature. Although most of the methods employed, earlier are now obsolete we shall discuss them briefly to show how consistently the research was carried out to improve and evolve better techniques.

1. **Mechanical Lever Method** : In this method, the experimental rod is fixed at one end A. The free end of this rod B is in contact with a vertical rod carrying a telescope pivoted at 0 and focused on a distant scale 'S' as shown in Fig. 2.1. Laplace and Lavoisier measured the increase in length (with proportionally amplified) on the scale. The friction to increase in length was avoided by using rollers r. Temperatures were maintained by a constant temperature bath α is then given by BB'/AB($t_2 - t_1$).

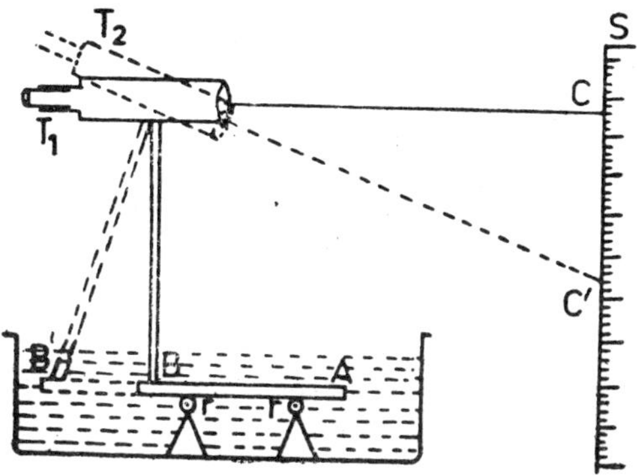

Fig. 2.1

2. **Optical Lever Method :** Here mechanical lever is replaced by optical lever. The experimental rod tilts the mirror M which in turn reflects the ray through twice the angle. Hence, the accuracy of the method is greatly increased. The method can be designed in several different ways one of them is as under:

Mirror At is mounted on a triangular metal framework with [three pointed legs whose feet are in one plane at the three corners of isosceles triangle. One leg of this frame is on the free end of the experimental rod

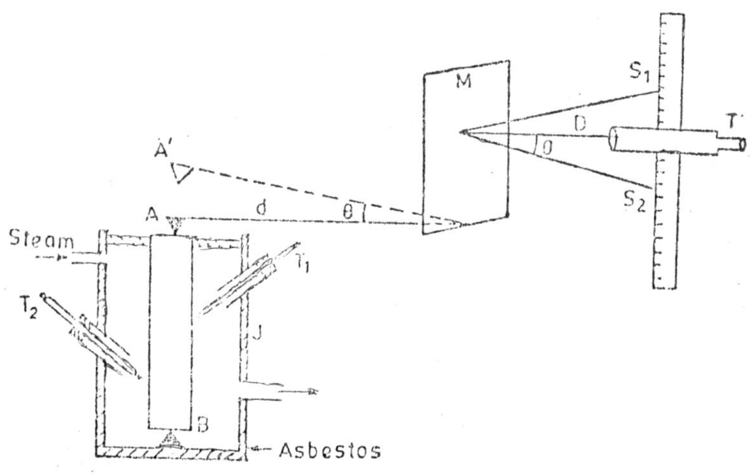

Fig. 2.2

vertically held inside the jacket. The rod, upon heating, expands only in upward direction. The mirror tilts due to this expansion which is reflected in the change in telescope settings as shown in the Fig. 2.2. Angle of tilt is determined by $\frac{AA'}{d} \cong \frac{S_1S_2}{2D}$

$$\therefore \quad AA' = \frac{d}{2D} \cdot S_1S_2$$

From which $\quad \alpha = \frac{d}{2D} \cdot \frac{S_1S_2}{AB}$

This method finds good use as a laboratory method for determination

3. **The Micrometer Screw Method :** In this method, the increase in the length of the rod is measured directly by means of a micrometer screw. In the Pullinger form, the spherometer is arranged at the top with its screw immediately above the free end of the experimental rod as shown in Fig. 2.3. The method is not an accurate one as it may accompany spherometric errors.

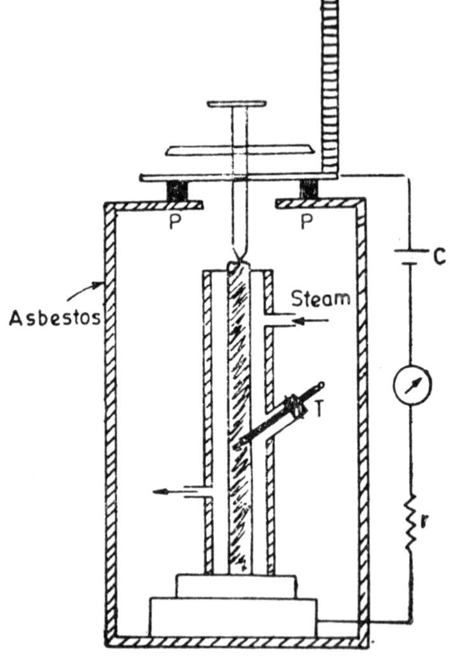

Fig. 2.3

4. **Comparator Method** : This is both a direct and a standard precision method and is used to compare standards of lengths at the International Bureau of Weights and Measures at Sevres (Paris). It compares a length between two marks on a given bar with the accurately known length between similar marks on a standard bar maintained at 0°C throughout.

The experimental bar AB is mounted horizontally on rollers in a double walled trough provided with glass windows. The bar is thus free to expand either way. In exactly similar manner is mounted a standard bar with two fine scratch marks on it, near its two ends, exactly a metre apart at 0°C, in another similar trough. The two troughs are mounted on a truck (on the rails) such that the two bars lie parallel to each other and can be brought into the field of view of two vertical microscopes. The microscopes remain unaffected (as they are mounted on rigid pillars of large thermal capacity) during the bar is heated up in a trough.

Two fine scratches X_1 and X_2, the fiducial marks, are made on the experimental bar near its ends at about the same distance as on the standard bar. The two troughs are now filled with melting ice (continuously stirred). Two carefully standardised thermometers indicate the temperatures (Fig. 2.4.).

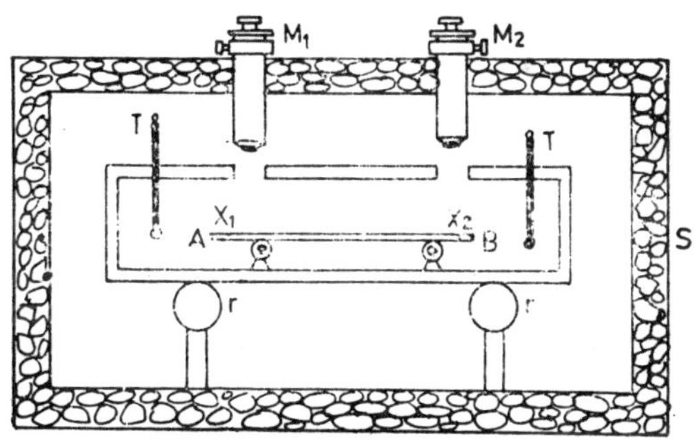

Fig. 2.4

When the bars attain the temperature 0°C, the trough containing standard bar is brought underneath the microscopes and their position is adjusted by means of the micrometer screws until their vertical cross-wires coincide with the images of the two fiducial marks on the bar. The distance between this (now exactly one metre) is carefully noted.

The standard bar is then wheeled away and the experimental bar is brought into position under the two microscopes. Again, their vertical cross-wires are made to coincide with the images of the fiducial points on it by shifting the position of the microscopes if necessary and again the micrometer readings are taken. The length l_0 of the experimental bar is thus obtained as 1 metre + or – total shift of the two microscopes, in true sense.

The temperature of the trough containing experimental rod is now raised by circulating hot water and maintained at a desired value by proper thermostatic controls. When bar attains a constant temperature t and expansion is completed, the microscopes are shifted outwards so that the vertical cross-wires again coincide with the images of fiducial marks on the rod. If e is the total outward shift of the microscopes, it is the increase in length of the bar so that the coefficient of linear expansion of the material is

$$\alpha = \frac{e}{l_0 t}$$

It is sometimes useful (to affect correction due to expansion changes in the position of the stone pillar) to bring the standard bar (kept still in the melting ice) once, again under the two microscopes and to focus them on to its two fiducial marks just to see if the micrometer readings for it are still the same.

Notes :

(i) By maintaining the experimental bar at different constant temperatures, we can study its expansion over different ranges of temperatures.

(ii) If specimen be in the form of tube, the temperature baths are not needed and water at a desired temperature can directly be passed through it.

(iii) If the wire specimen is suspended from a rigid support and a small weight suspended at the free end (to prevent sagging), the microscopes are replaced by cathetometers.

5. Hennings' Method : It is a simple and ingenious method for accurate determination of the relative or differential expansion of a bar of a given material with respect to that of a standard substance like fused silica or Jena Glass, whose expansion is small and accurately known over the desired range of temperature.

Henning has thus combined the advantages of both the comparator and the micrometer methods in his method.

The experimental rod R, with its two ends ground to perfect planes, about 50 cm long rests on a pointed base vertically inside a long fused silica or Jena glass tube. On the top of the rod (the free end) rests a pointed end of another rod of the material of the tube as shown in Fig. 2.5. Scales S'. S and V are engraved on the upper part of the tube and on the rod. The whole apparatus is immersed up to half length of the rod S, first in a cold and then in a hot, constant temperature bath and the relative shift of the two scales is measured by means of a microscope. This gives the expansion of the experimental rod, relative to silica tube of equal length. It being assumed that the temperature at each point on the silica rod is the same as that of the tube at the same level.

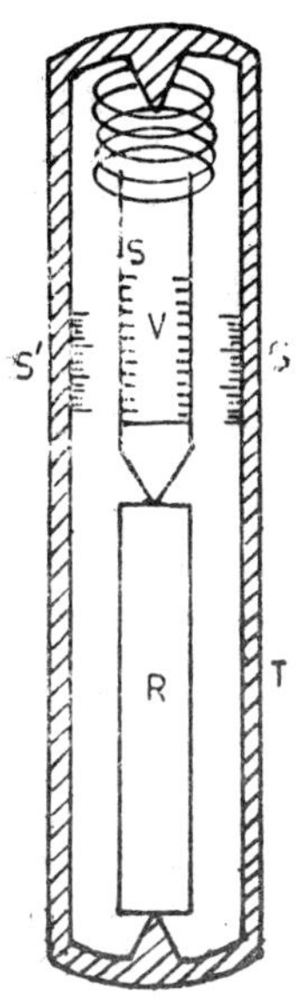

Fig. 2.5

To obtain the absolute expansion of the material of the rod R, the expansion of the tube material must be accurately determined in view of its small but definite expansion coefficient ($\sim 5 \times 10^{-5}$ m/°C). The advantages of the fused silica tube are (i) it can be used over a wide range of temperatures, (ii) its coefficient of linear expansion is small.

6. **Interference Method** : None of the methods, discussed above, is suitable for the determination of the coefficient of expansion of a material, available in a small quantity/size as in the case of a crystal due to the difficulty in measuring accurately small increase in length involved.

Fizeau in 1864, devised a suitable method for measuring small expansions using the wavelength of light as his standard or unit for small increases in small-sized specimen. By basing his method on the interference of light he could measure expansions as small as 6×10^{-8}m.

The specimen is taken in the form of a plate of a known thickness (to be taken as-length in the expansion measurement), with its upper and lower surfaces ground optically plane and parallel to each other. It is placed on a horizontal, polished metal plate AB, supported on three levelling screws .as shown in Fig 2.6 A flat, parallel sided, optically worked .glass plate CD supported on the levelling screws is adjusted such that it lies quite close and almost parallel to the crystal plate. A thin wedge-shaped film of air is thus enclosed in between the two, the angle of wedge being very small.

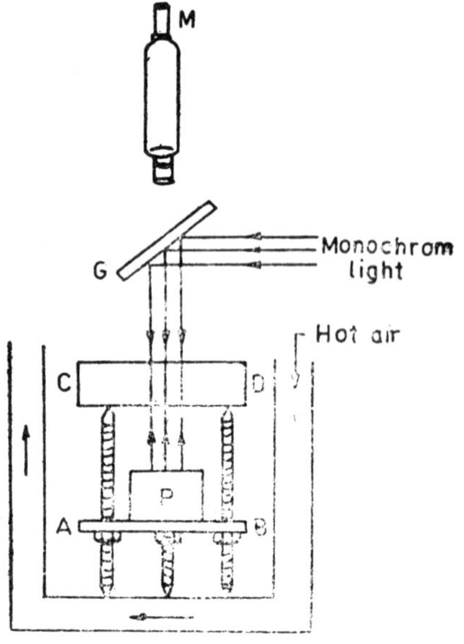

Fig. 2.6

Whole apparatus is arranged inside a double-walled chamber which can be heated either by circulating hot air through annular space between its walls, or directly with thermostatic control of its temperature.

A parallel beam of monochromatic light (from sodium, cadmium or mercury lamp) is allowed to fall normally on plate CD, after reflection at a glass plate G inclined at .an angle 45°. Part of the beam is reflected from the lower surface of CD and the rest traverses the air film and suffers reflection at the upper surface of the crystal plate P. A path difference is thus introduced between two parts of the beam giving rise to straight, parallel and equi-spaced fringes on the surface of the film itself (the colour of the bright fringes being the same as that of the monochromatic light used). These are observed through a low power microscope M arranged vertically above and one of the dark fringes is focused on the cross-wire.

The chamber is now gradually heated. The crystal expands (so also the levelling screws). The thickness of the air film changes by an amount equal to the difference between these two expansions. This results in the shift of fringes which appear to successively jump across the cross-wire. The number of dark fringes thus crossing the cross-wire is carefully counted until their displacement ceases (*i.e..*, until the temperature of the chamber is constant).

Then if Y metres be the changes in the thickness of the air film and N fringes cross the cross-wire, we have

$$Y = \frac{N\lambda}{2} \qquad (\lambda \text{ is wavelength of light used})$$

This change in the thickness of the air film is the net result of (/) the expansion of the crystal which decreases the thickness of the film by reducing the distance between CD and itself, and (+) the expansion of the levelling screws which increases the thickness of the film by pushing plate CD upwards. Thus to obtain the change in the thickness of the film due to expansion of the crystal alone, we must subtract from y the change in the thickness due to expansion in levelling screws.

For this purpose, we remove the crystal, bring the plate CD closer to AB and repeat the experiment within the same range of temperature as before. The interference now occurs between the rays reflected from the lower surface of CD and the upper surface of AB. If now N' is the number of fringes that cross the field of view then the change in the thickness of air film due to expansion of the levelling screws = N' $\lambda/2$.

∴ Change in thickness of the air film due to expansion of the crystal $= \frac{N\lambda}{2} - \frac{N'\lambda}{2}$; and the coefficient of linear expansion of the crystal

$$\alpha = \frac{(N - N')\lambda/2}{l(t_2 - t_1)} = \frac{(N - N')\lambda}{2l(t_2 - t_1)}$$

where l is the thickness of the crystal and $(t_2 - t_1)$ is the rise in temperature for the experiment.

A careful investigator can detect a shift 1/5 of a fringe. As λ is of the order 6×10^{-5} cm, a change in the thickness as small as 6×10^{-5} cm. $\equiv 6 \times 10^{-7}$ m can be accurately measured.

Tutton devised a simple method to eliminate altogether the effect of expansion of the levelling screws by means of an aluminium compensator. The compensator comp. is just a three-legged table of aluminium (Fig. 2.7), which is placed on the polished plate AB. The crystal P is placed on this table. The thickness of the aluminium plate (length of the legs of the table) is so adjusted that its expansion just balances the expansion of the levelling screws over a wide range of temperatures. So we have to note, only once, the shift of dark fringes.

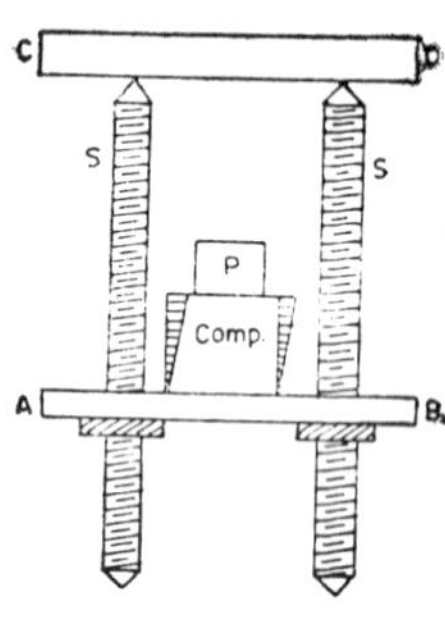

Fig. 2.7

2.8 OTHER MODIFICATIONS TO FIZEAU'S METHOD

(i) **Abbe's Modification :** Here a plano-convex lens of quartz is used in place of the glass plate CD. A circular quartz ring forms the base to this lens and surrounds the crystal P as shown in Fig. 2.8. When the chamber is heated this thickness of the film changes and Newton's rings (formed in this case) appear to expand outwards. Here again the cross-wire of the eye piece set on one of the dark rings records a number of crossing it as the temperature of the crystal rises from t_1 to t_2°C. Then, as before, the change in the thickness of the air film due to expansion of the crystal plus the quartz ring is equal to $N\lambda/2$. We neglect here the linear expansion of fused quartz ring which is almost zero. But for accurate work the expansion of the quartz ring can be eliminated as in the Fizeau's method.

(ii) **Abbe and Pulfrich improvement :** These experimenters used both plates AB and CD of quartz and used quartz ring also to support the plate CD. The experimental crystal was cut in the form of a cylinder and placed inside the quartz ring. In this case the interference fringes were straight and parallel. This method has been successfully used to record the expansion coefficient of non-crystalline materials also.

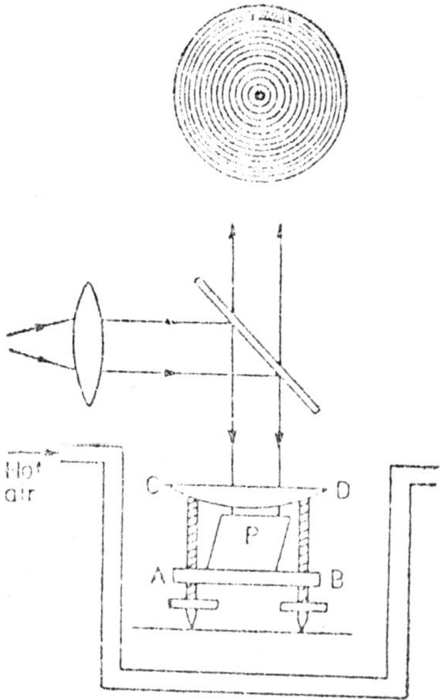

Fig. 2.8

(iii) Air-wedge Dilatometer Method : Priest, in the year 1920, devised method in which the expansion of the crystal is deduced from a change the width of the interference fringes instead of their displacement.

The essentials of his apparatus are shown in Fig. 2.9. AD is an optically plane polished glass plate hinged at A on the horizontal base plate AB. Plate AB is also polished, optically plane and has a sharp knife edge at A. The other end of the cover plate AD rests on the pointed upper tip of the crystal. The specimen material is taken in the form of a small cylinder and is held vertically so that an air wedge of a small thickness 0.1 to 0.3 mm is formed between the cover plate and the glass plate (shown greatly exaggerated in the figure).

A parallel beam of monochromatic light is allowed to fall on the cover plate. The interference pattern (straight and parallel bright and dark fringes) is observed through the microscope M. The number of fringes between two fine lines X_1 and X_2 (drawn on the plate AD by a fine diamond point) is counted carefully. The thickness of each fringe, say δ_1 is thus evaluated.

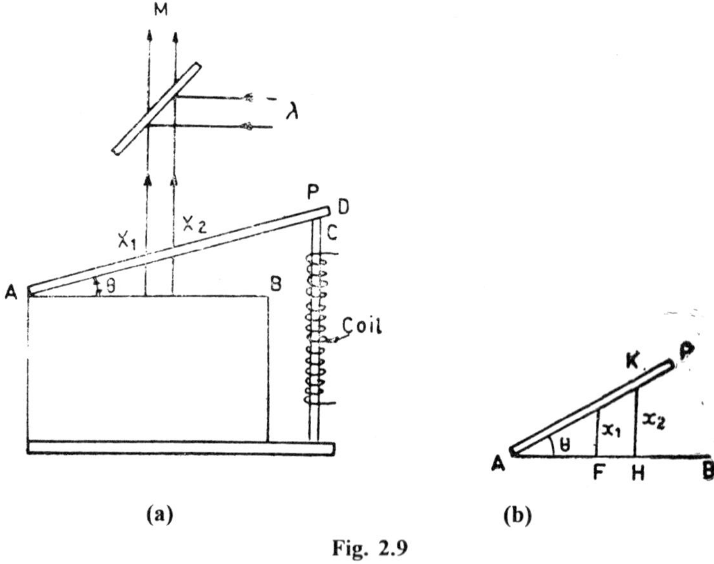

Fig. 2.9

The crystal is now heated to a desired temperature by passing a current through the heating coil (wrapped around it) and the rise in temperature is noted on a sensitive thermocouple. Due to linear expansion of the crystal the plate is tilted slightly more upwards whence thickness of the air film wedge increases. This results in decrease in fringe width. A larger number of fringes are there, now, between X_1 and X_2. The fringe width is again calculated, let it is δ_2.

Consider a small portion of the air film. Let the thickness of the film at F be x_1 and at H be x_2. Such that FH = δ_1 the width of the fringe at initial temperature.

Now as we know, for one dark band another succeeds when thickness of the film increases by $\lambda/2$, where λ is wavelength of monochromatic light used.

It follows, therefore, that

$$x_2 - x_1 = \lambda/2.$$

Then, if θ be the angle of ,the air wedge (Fig. 2.10) and is small enough, we have

$$\theta = \sin\theta = \tan\theta$$

$$\text{or,} \qquad = \frac{x_1}{AF} = \frac{x_2}{AH}$$

$$\therefore \qquad \theta = \frac{x_2 - x_1}{AH - AF} = \frac{x_2 - x_1}{FH} = \frac{\lambda}{2\delta_1}$$

but θ is also t_1/D where t_1 is the thickness of the air film at point P where the cover plate rests on the crystal tip and D = AB = AP

$$\therefore \qquad \frac{t_1}{D} = \frac{\lambda}{2\delta_1} \qquad \text{or,} \quad t_1 = \frac{D\lambda}{2\delta_1}$$

A similar relation is obtained for the expanded crystal with its new position. Let the thickness at P of the air film be now t_2 then

$$t_2 = \frac{D\lambda}{2\delta_2}$$

$$\therefore \text{ Expansion in crystal } = t_2 - t_1 = \frac{D\lambda}{2}\left[\frac{1}{\delta_2} - \frac{1}{\delta_1}\right]$$

Thus, knowing the increase in length of the crystal its original length and the rise in temperature, we can calculate its coefficient of linear expansion.

2.9 VARIATION OF THE COEFFICIENT OF EXPANSION WITH TEMPERATURE, GRUNEISEN'S LAW

The coefficient of linear expansion, a is not really a constant quantity, *i.e.*, it is not strictly a linear function of temperature. It is found that it decreases with decrease in temperature and tends to become zero at absolute zero of temperature. Exactly in the same manner as the specific heat of a solid at constant pressure, C, decreases with decrease in temperature. Gruneisen showed that the ratio α/C_p was constant for a particular isotropic solid. This is the Gruneisen law stated as "For an isotropic solid, the ratio of its coefficient of linear expansion to its specific heat at constant pressure, is constant at all temperatures.

2.10 LINEAR COEFFICIENT OF EXPANSION OF CRYSTALS BY X-RAY METHODS

The atoms or ions in a crystal are arranged in a regular pattern in space. A crystal thus can be conceived as composed of a series of parallel planes, the distance between these planes being d, the lattic spacing.

Let such a three-dimensional crystal grating reflect X-rays at an angle θ, then

$$n\lambda = 2d \sin \theta$$

Now, it is found that d increases with temperature and can be represented by the formula

$$d = d_0 (1 + \alpha t)$$

which when substituted in the X-ray formula agrees experimentally quite well (Here λ is the wavelength of X-rays). Hence λ can be determined from X-ray analysis.

Expansion of Anisotropic Solids

An anisotropic or non-isotropic (*i.e.*, crystalline) solid on being heated expands differently in different directions. This effect was first observed by Mitscherlich who observed the variation of the dihedral angles of crystals of Iceland spar with change of temperature (dihedral angle is the angle between faces of cleavage).

In the case of most of such crystals we have three mutually perpendicular directions such that if a cube is cut out of the crystal with its sides parallel to these directions the angle will remain the same even when heated. But the linear expansion along the three edges may be different. These three directions are called the three principal axes of dilation. It is, however, not necessary that the principal axes of dilation are identical to the axes of symmetry. A knowledge of these enables one to calculate linear expansion of the crystal in any other direction.

2.11 THE COEFFICIENT OF CUBICAL EXPANSION OF A CRYSTAL

Let us consider a cube, cut out of a crystal, with its edges parallel to the three principal axes of dilation. Let l_0 be the length of each edge at 0°C and that its volume be V_0. Further, let the three principal coefficients of expansion be α_x, α_y and α_z. Then on heating it through t°C the crystal (cubic) will become a parallelepiped of edges $l_0(1 + \alpha_x t)$, $l_0(1 + \alpha_y t)$, $l_0(1 + \alpha_z t)$ so that the new volume

$$V = l_0(1 + \alpha_x t)\, l_0(1 + \alpha_y t)\, l_0(1 + \alpha_z t)$$
$$= l_0^3\{1 + (\alpha_x + \alpha_y + \alpha_z)\, t + ...\}$$

but $V_0 = l_0^3$

$\therefore$ $V \cong V_0(1 + [\alpha_x + \alpha_y + \alpha_z) t)$

or $V \cong V_0(1 + \gamma t)$.

Thus, the coefficient of cubical expansion for an anisotropic crystal is approximately equal to the sum of its three principal coefficients of dilation.

There are three different cases:

(i) In the case of crystals like rock-salt, diamond, etc. which belong to the cubic system of crystals, the three principal coefficients of expansion are equal to each other, *i.e.*, $\alpha_x = \alpha_y = \alpha_z = \alpha$ $\therefore$ $\gamma = 3\alpha$ in such a case cubical expansion coefficients of isotropic and anisotropic solids is the same.

(ii) In the case of uniaxial crystals (crystals which contain only one optic axis), like those of Iceland spar and emerald, there is an axis of crystalline symmetry, perpendicular to which their physical properties are the same in all directions. Hence if α_x be the coefficient of linear expansion along the axis of symmetry (which is one of the principal axes), then the principal coefficients of linear expansion in the other two directions α_y and α_z will be equal

$\therefore$ $$\gamma = \alpha_x + \alpha_y + \alpha_z + \alpha_x + 2\alpha_y$$

In other words, we have only two coefficients of linear expansion in such systems.

(iii) In the case of crystals which neither belong to the cubic nor to the uniaxial system, all the three coefficients of linear expansion are different from one another.

2.12 COEFFICIENT OF LINEAR EXPANSION IN A DIRECTION OTHER THAN THE PRINCIPAL AXES

Let OX, OY and OZ be the three principal axes of a crystal and let we are required to determine the value of the coefficient of linear expansion α along OP making an angle β, δ and v with the three axes respectively as shown in Fig. 2.10.

Let x, y, z be the coordinates of a point P in the crystal and let OP = r. The at initial (0°C) temperature

$$x^2 + y^2 + z^2 = r^2$$

Now, let the crystal be heated to a temperature, t°C so that the point P moves to P' and its new coordinates are $x(1 + \alpha_x t)$, $y(1 + \alpha_y t)$ and $z(1 + \alpha_z t)$, and, therefore OP' = r' is such that

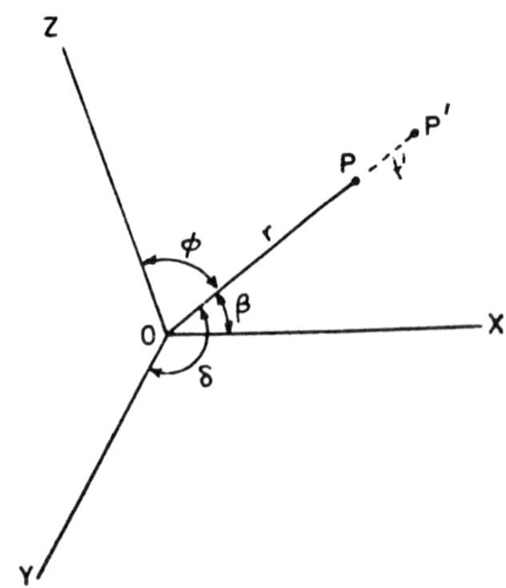

Fig. 2.10

$$r'^2 = x^2(1 + \alpha_x t)^2 + y^2(1 + \alpha_y t)^2 + z^2(1 + \alpha_z t)^2$$

$$\cong x^2 + y^2 + z^2 + 2t(x^2\alpha_x + y^2\alpha_y + z^2\alpha_z) + ...$$

$$\cong r^2 + 2t(x^2\alpha_x + y^2\alpha_y + z^2\alpha_z) + ...$$

$$\therefore \quad r'^2 - r^2 \cong 2t(x^2\alpha_x + y^2\alpha_y + z^2\alpha_z)$$

$$\therefore \quad r' = r \cong \frac{2t(x^2\alpha_x + y^2\alpha_y + z^2\alpha_z)}{(r'+r)}$$

$$\therefore \quad (r' = r) \cong \frac{t}{r}(x^2\alpha_x + y^2\alpha_y + z^2\alpha_z)$$

where (i) we have neglected higher powers of α, (ii) we wrote, $(r' - r)(r' + r) = r'^2 - r^2$, and (iii) also use $r' \cong r$ or $r' + r = 2r$ since r' is not much different from r.

$$\therefore \quad \alpha = \frac{r^1 = r)}{rt} = \frac{1}{r^2}(x^2\alpha_x + y^2\alpha_y + z^2\alpha_z)$$

i.e.,

$$\alpha = \left(\frac{x^2}{r^2}\alpha_x + \frac{y^2}{r^2}\alpha_y + \frac{z^2}{r^2}z_z\right)$$

or, $\alpha = \alpha_x \cos^2\beta + \alpha_y \cos^2\delta + \alpha_z \cos^2\varphi$

as $x/r = \cos\beta$, $y/r = \cos\delta$ and $z/r = \cos\varphi$

Thus, the coefficient of the linear expansion of the crystal in any direction can be obtained in terms of its principal coefficients of linear expansions and the direction cosines.

Important Deduction : We notice from above theory that

(i) In a direction for which $x^2\alpha_x = y^2\alpha_y + z^2\alpha_z$ is zero there will be no expansion or contraction in the crystal.

(ii) If the direction OP be equally inclined to the three principal axes (*i.e.*, $\beta = \delta = \varphi$), we have, $\cos^2\beta = \cos^2\delta = \cos^2\varphi = 1/3$

or, $$\alpha = \frac{1}{3}(\alpha_x + \alpha_y + \alpha_z) = \frac{\gamma}{3}$$

or, $$\gamma = 3\alpha.$$

Thus, for this direction, we can obtain γ simply from the value of α. (iii) If we choose α_1, α_2, α_3 such that these directions are mutually perpendicular then

$$\alpha_1 = \alpha_x \cos^2\beta_1 + \alpha_y \cos^2\delta_1 + \alpha_z \cos^2\varphi_1$$

$$\alpha_2 = \alpha_x \cos^3\beta_2 + \alpha_y \cos^2\delta_2 + \alpha_2 \cos^2\varphi_2$$

$$\alpha_3 = \alpha_x \cos^2\beta_3 + \alpha_y \cos^2\delta_3 + \alpha_2 \cos^2\varphi_3$$

or, $$\alpha_1 + \alpha_2 + \alpha_3 = \alpha_x (\cos^2\beta_1 + \cos^2\beta_2 + \cos^2\beta_3)$$
$$+ \alpha_y(\cos^2\delta_1 + \cos^2\delta_2 + \cos^2\delta_3)$$
$$+ \alpha_z(\cos^2\varphi_1 + \cos^2\varphi_2 + \cos^2\varphi_3)$$

$$\therefore \alpha_1 + \alpha_2 + \alpha_3 = \alpha_x + \alpha_y + \alpha_z \cong \gamma$$

$$(\because \cos^2\beta_1 + \cos^2\beta_2 + \cos^2\beta_3 = 1, \text{ etc.})$$

Thus, the sum of the linear coefficients of a crystal, in three mutually perpendicular directions, is a constant and is equal to its cubical coefficient.

2.13 SOLIDS WITH LOW EXPANSION COEFFICIENT

Among solids diamond, fused quartz and fused silica have the lowest expansion coefficients. Quartz and silica, hence, has been of great practical utility. Fused quartz is used in the construction of large telescope mirrors. Silica-glass (quartz, fused and then resolidified into a non-crystalline state) is almost invariably used for making thermometer tubes, etc.

Silica vessels can be heated to quite high temperatures and can be cooled suddenly without a danger of craking as it has (i) very low (one-twentieth of ordinary glass) expansion coefficient ($\alpha = 5 \times 10^{-7}$) and (ii) a truly linear function of temperature right up to 1000°C with just a side band at either end. The expansion coefficient even assumes a negative value below –80°C.

Expansion of Alloys : Alloys often exhibit curious anomalies compared to the pure metals of which they are composed. Thus, for example,

$$\alpha_{zn} = 26 \times 10^{-6}/°C$$

$$\alpha_{Al} = 18 \times 10^{-6}/°C$$

but (95% Zn + 5% Al) has $\alpha = 28 \times 10^{-6}/°C$

Similarly,

$$\alpha_{F6} \cong \alpha_{Ni} = 12 \times 10^{-6}/°C$$

but the alloys of Fe and Ni (called the platinites) have $\alpha = 8 \times 10^{-6}/°C$. Dr. C.E. Guillaume showed the alloy of Ni–30%, Fe–64%. Mn–4%, C–0.1% and a little of silicon had $\alpha = 1 \times 10^{-6}/°C$ which can further be reduced to near zero after some treatments (such an alloy is invar). Still more recent discoveries of alloys are:

(a) Stainless invar—a less corrodible and non-expansible alloy of iron, cobalt and chromium, (b) elinvar-similar to invar but unaffected by the magnetic field. It is invar with more of chromium. It is special in that its elasticity is unaffected by changes in temperature and hence finds use in hair springs of balance wheels, construction of chronometers, etc.

2.14 SOME PRACTICAL APPLICATIONS OF THERMAL EXPANSION

It is well known that the surveyor's tape, substandard scales and clock pendulums suffer mainly from two defects (i) they have a higher magnetic susceptibility, and (ii) slow change in length with age (1 m rod is found to reduce by 0.023 mm in 22 years). Such and many other defects may be taken care of by alloying and making some other provisions based on the thermal expansion as indicated below which are due to (i) change in density, (ii) change in moment of inertia, and (iii) change in elasticity.

I. Cases Where Thermal Expansion is Helpful or of Practical Utility

1. **Fitting Iron Hoops on Cart Wheels.:** By heating the iron hoop until it expands enough to be slipped over the wooden wheel and subsequently cooled for firm grip.
2. **Strengthening and Preventing Bulging Walls from Collapsing:** Tie-rods of iron are passed through two opposite walls right across the room and heated in the centre. As they expand, iron plates or S-shaped iron strips are slipped on to their threaded ends and secured tightly in position with nuts and bolts as close to the outside surface as possible. On cooling, the tie-rods contract and pull the two walls inwards.
3. **Rivetting Boiler Plates and Fitting Joints:** To ensure steam-proof joints in boiler plates red hot rivets are fitted which fasten more tightly on cooling.
4. **Shrinkage Fits:** When two parts of a joint which are to be fitted together, the inner one is first cooled to –80°C (by solid CO_2). As it warms up it expands and forms a firm fit.
5. **Hot Wire Ammeter:** A current to be measured is passed through a wire under tension. The resistive heating causes expansion of the wire from which current is estimated
6. **Thermostats and Solid Expansion thermometer :** The thermostat (a device to control the temperature of a body) can be formed from the property of solid's expansion. This has been used to close and open the electrical circuit automatically. Fig. 2.11 gives a diagrammatic representation of a simple thermostat based on the use of bimetallic strip. It finds use in electric iron. As the name suggests, it consists of a strip of two different metals, brazed side by side, having different expansibilities, so that when heated it gets bent. This banding produced in the strip (S in the figure) due to different expansions of the two strips, pushes C up and thus the electrical circuit breaks down Later when strip cools 'C' again comes closer to D and finally allows the current to flow in the circuit.

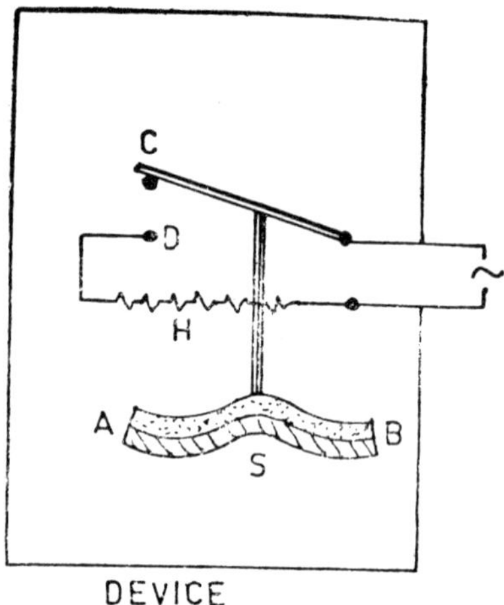

Fig. 2.11

The bending on heating of bimetallic strip has also been utilised in temperature measurement.

II. Cases Where Thermal Expansion is Hindrance

1. **Expansion of Metal Scales :** Measuring scales on metal rod, etc., are correct only at temperature at which the marks are engraved on them. At higher temperatures distance between two successive marks on it increases, and so true reading has to be

   $H = H_t/(1 + \alpha t) \cong H_t(1 - \alpha t)$

   where H and H_t are the readings on ideal scale and on metal scale at t°C respectively.

2. **Laying of Railway Tracks :** and construction of suspension bridges where allowance is made for expansion of metal mass.

3. **The Compensated Pendulum :** The pendulum clocks' time period depends on its pendulum length l, $T \propto \sqrt{l}$, *i.e.*, the time period of the pendulum depends linearly on square root of its length. The length alters with temperature hence the clock may give incorrect time unless it has a compensated pendulum (whose length remains unaltered by changes in temperature) like:

   (i) Harrison's grid iron compensated pendulum : Harrison used two different metals with different expansibilities and

annulled, expansion by expansion to attain compensation. The pendulum consists of five steel and four brass rods arranged alternately as shown in Fig. 2.12. The bulb is attached to the central steel rod. The steel rods are free to expand downwards while the brass rods, upwards. The two expansions balance each other at all temperatures, if the length of the rods are adjusted according to

$$l_{iron}\ \alpha_{iron}.t = l_{brass}\ \alpha_{brass}.t$$

or $$l_{iron}/l_{brass} = \alpha_{brass}/\alpha_{iron} \cong 3 : 2.$$

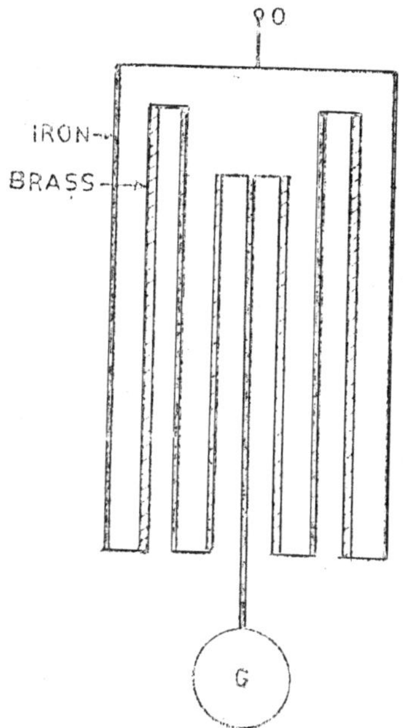

Fig. 2.12

(ii) *Graham's mercury pendulum* : Graham introduced a pendulum that carried mercury in the glass cylinder as shown in the Fig 2.13. Here compensation is affected by the upward expansion of mercury, off setting the downward expansion of the steel rods. From calculations it is found that mercury column should be 0 145 times the length of steel rod.

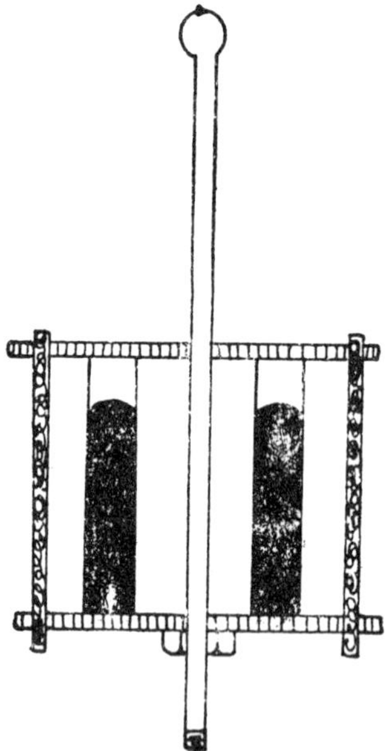

Fig. 2.13

(4) **The Compensated Balance Wheel o a Watch :** The rate of running of a watch or a chronometer is controlled by its balance wheel which is a sort of torsional pendulum, oscillating under the elastic control of a spiral hair spring. Its period of oscillation is given by $T = 2\pi\sqrt{1/c}$ where I is moment of inertia of the wheel about its axis of oscillation and C, the restoring couple per unit twist of the hair spring.

With the increase in temperature, two distinct changes occur:

(*i*) *Increase in the size of the wheel causing an increase in I, and*

(*ii*) *The decrease in the elasticity of the spring causing a decrease in the value of C.*

Both changes contribute to an increase in the period of oscillation. The two changes, thus, are to fee annulled. First by not allowing the radius of the wheel to increase, and second by decrease in the radius of the wheel with the rise in temperature. The two changes together are compensated by a decrease in the radius of the wheel with rise in temperature so that I/C remains the same. This has been achieved by dividing the rim of the wheel into segments. The segments are of bimetallic strips with one end of each connected to centre by a spoke as shown in the Fig. 2.14. The more expansible material is on the outer side and the less expansible on the inner side. Thus higher temperatures will reduce the radius. There are also screw-weights on the rim to adjust the time period so also for loss in elasticity (lowering) of the spring by any other effect also. The invention of invar has, however, simplified the mechanism as no compen-sator is needed.

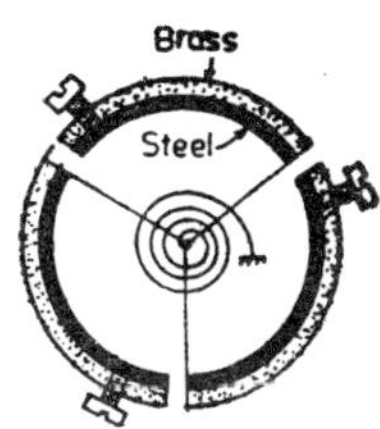

Fig. 2.14

(5) Expansion in Optical and Electrical Apparatus : To have the images very accurately figured, the mirrors of reflecting telescopes should have low expansion and high thermal conductivity (which cannot be met in one material). Thus, mirror surface distortion may be attended by rise in temperature. Thus, here we need to use fused silica and pyrex.

Similarly, as electrical constants of inductances and condensers also depend on their dimensions, temperature causes change in these electrical quantities. Thus great ingenuity is needed to maintain the quantities same by compensation.

(6) Glass Metal Sealing : Many appliances (such as discharge tube) need conductor wires sealed in the glass apparatus, there also one has to be careful to match the expansion coefficients of the two (invar steel, molybdenum, platinum wire, etc., may be used for fused silica) or use the soft metal which can yield under the stress produced due to changes in temperature.

(7) Variation of the Frequency of a Tuning Fork with Temperature : The frequency of a tuning fork is given by

$$n \propto \frac{K}{l} \sqrt{E/P}$$

where K is radius of gyration, l is length, E Young's modulus and P is the density of tuning fork material.

Now, as K varies directly as l while P varies inversely as l_3 we have

$$n \propto \sqrt{El} \qquad \text{or} \qquad n = A\sqrt{El}$$

where A is a constant. In this equation, we have two variables for temperature changes, *viz,*

$$E_t = E_0(1 - \varphi t) \text{ and } l_t = l_0\,(1 + \alpha t)$$

so that $$n_0 = A\sqrt{E_0 l_0}$$

and $$n_t = A\sqrt{E_1 l_1}$$

where subscript o is for quantities at 0°C and t for quantities at t°C, then

$$n_t = A\sqrt{E_0(1 - \varphi t)l_0(\varphi - \alpha t}$$

$$A\sqrt{E_0 l_0[1 - t(\varphi - \alpha) + \ldots]}$$

$$\cong n_0[1 - t(\varphi - \alpha)]^{1/2}$$

i.e., $$n_t \cong n_0\left[1 - \frac{1}{2}(\varphi - \alpha)t\right].$$

This equation enables one to find the temperature coefficient of Young's modulus for the solid as has been suggested by Koenig. He found that φ was approximately twenty times as great as α

SECTION II: EXPANSION OF LIQUIDS

2.15 VOLUME EXPANSION COEFFICIENTS

The liquids are incompressible and possess a definite volume but they differ from solids in that (i) they have no shape of their own, (ii) they are not generally anisotropic, and (iii) their expansion is several times larger that the solids.

To study expansion of liquids it must be noted that (i) liquids need container and (ii) liquids exhibit a marked difference in their rates of expansion over different ranges of temperatures.

In view of the uneven expansion of liquids over different ranges of temperature, it is usual to define the coefficient of expansion of liquids in two different ways, *viz.,* (i) as its mean coefficient of expansion over a given range of temperature, and (ii) as its zero coefficient of expansion in the range 0°C to 1°C.

(i) *Mean coefficient of expansion of a liquid,* γ_m, for a given ranger of temperature is the increase in volume per unit volume of the liquid per degree Celsius rise of temperature within this range. *i.e.*,

$$\gamma_m = \frac{V_2 - V_1}{V_1(t_2 - t_1)}$$

or simply

$$(\gamma_m)_{t_1 \to t_2} = \frac{V_1 - V}{V \cdot (t_2 - t_1)}$$

i.e., $$V' = V[1+\gamma_m (t_2 - t_1)]$$

which is also written as

$$V = V_0 (1 + \gamma_m t)$$

if t_2, = t (and t_1 = 0. Here t is temperature and V is corresponding volume of the liquid.

(ii) Zero coefficient of expansion of a liquid, γ_0 is the increase in volume per unit volume of the liquid when heated from 0°C to 1°C.

i.e., $$\gamma_0 = \frac{V_1 - V_0}{V_0}$$

If the temperature range is very small. *i.e.*, $(t_2 - t_1) = \Delta t$ and tends, to zero, we have

$$\gamma_m = \gamma_0 = \gamma = \frac{dV}{V \cdot dt}$$

now this γ is true or the absolute coefficient of the liquid. Rearranging the equation

$$\frac{dV}{V} = \gamma dt$$

or $\log V = \gamma t + C$

Since $V = V_0$

when $t = 0$

we have $C = \log V_0$

so that $\log V - \log V_0 = \gamma t$

or $V = V_0 e^{\gamma t}$.

2.16 APPARENT AND REAL EXPANSION OF A LIQUID

A liquid has to be in a container which too expands along with the liquid when, heated (though to a small extent). As a result the liquid as it expands has also to fill the increased volume of the container. Thus the observed or apparent expansion is less than its real or absolute expansion by an amount equal to the cubical expansion of the container.

The coefficient of apparent expansion of a liquid is thus defined as its apparent fractional increase in volume per degree (usually celsius) rise of temperature and is denoted as

$$\gamma a \, {}^\circ C^{-1}$$

or $\gamma a K^{-1}$.

Similarly, the *coefficient of real or absolute expansion of the liquid* is defined as its real fractional increase in volume per degree rise in temperature and is denoted as

$$\gamma r \, {}^\circ C^{-1}$$

or $\gamma r K^{-1}$.

2.17 RELATION BETWEEN γ_a AND γ_r

Suppose we have a liquid in a glass container occupying a volume V_0 at 0°C. Let the vessel be heated to t°C (without liquid) the volume of the vessel would be

$$V = V_0(1 + \gamma_g t)$$

and the liquid will have volumes

$$Vr = V_0(1 + \gamma_r t)$$

$$V_a = V_0(1 + \gamma_a t)$$

for its real and apparent expansions respectively.

Here obviously V_a is the volume of the container occupied by the liquid at t°C. Since graduations on the container correspond to the true volume only at 0°C hence it is so for the liquid also at 0°C only. So the real volume of the liquid must be, in terms of graduations,

$$V_r = V_a(1 + \gamma_g t)$$

$$V_r = V_0(1 + \gamma_a t)\,(1 + \gamma_g t)$$

or $V_0(1 + \gamma_r t) = V_0(1 + \gamma_a t)\,(1 + \gamma_g t)$

i.e. $1 + \gamma_r t \cong 1 + (\gamma_a + \gamma_g)t + A_a\gamma_g t^2$

$\therefore$ $\gamma_r = \gamma_a + \gamma_g + \gamma_a\gamma_g t$

or, $\gamma_r \cong \gamma_a + \gamma_g$

i.e., the coefficient of absolute expansion of a liquid is approximately equal "to the sum of its coefficient of apparent expansion and the coefficient of cubical expansion of the containing vessel.

2.18 RELATION BETWEEN VOLUME AND DENSITY OF A LIQUID

Suppose mass m of a liquid occupies, a volume V_0 at 0°C. Then if P_0 and δ_t be its densities at 0°C and t°C respectively and γ_m, its mean coefficient of expansion for a given range of temperature

We have

$$P_t = \frac{m}{V_t} = \frac{m}{V_0(1 + \gamma_m t)} = \frac{P_0}{(1 + \gamma_m t)}$$

$\therefore$ $P_0 = p_t(1 + \gamma_m t)$

or $p_t = P_0(1 - \gamma_m t)$

then for two temperatures t_1 and t_2, if γ_m is mean coefficient, we have

$$\rho_2 = \rho_0(1 + V_m t_2),$$

$$\rho_1 = \rho_0(1 + V_m t_1),$$

$\therefore$ $\rho_2 - \rho_1 = -\rho_9 V_m(t_2 - t_1),$

or, $$\gamma_m = \frac{\rho_1 - \rho_2}{\rho_0(t_2 - t_1)}$$

or also $\rho_1 = \rho_2[1 + \gamma_m(t_2 - t_1)]$

so that $\rho_2 = \rho_1[1 - \gamma_m(t_2 - t_1)]$

or, $$\gamma_m = \frac{\rho_1 - \rho_2}{\rho_1(t_2 - t_1)}$$

Thus, from the densities of the liquid its mean coefficient of absolute expansion can be determined which is quite unaffected by the expansion of the container. The two equations denote if $t_2 > 0 \not> t_1$, and also from the fact that $\gamma \neq \gamma_m$. Hence, it is always good to take $t_2 - t_1$ as small possible and find γ_m from

$$\gamma_m = \frac{\rho_1 - \rho_2}{\rho_1(t_2 - t_2)} \quad ...(i)$$

2.19 DETERMINATION OF EXPANSION COEFFICIENTS OF LIQUIDS

It is difficult to find absolute expansion coefficient of the liquid. The only direct method can be that suggested in Section 2.18 equation (i) *i.e.*, comparing the densities of the given liquid at two different temperatures. This is the method of balancing a cold column of a liquid against a hot column of the same liquid (we can balance two different liquids also with same density calculation but it is always good to standardise a single ideal liquid and then obtain the data for others. A number Of workers did this job one after another on mercury). Once the coefficient of absolute expansion of the liquid (mercury) was known the problem of the coefficient of absolute expansion of other liquids is solved. At first, coefficient of apparent expansion of mercury in any vessel is determined. This value is then subtracted from its coefficient of absolute expansion. This difference corresponds to the cubical expansion of the vessel. Now the coefficient of apparent expansion of the given liquid is determined in that vessel. Then added cubical expansion of the vessel to this apparent value to get absolute coefficient of expansion. We will discuss some of the methods used for determining expansion coefficients.

I. Determination of Coefficient of Absolute Expansion of a Liquid

(i) Balancing Column Method : In this method we balance, in a U-tube (Hare's apparatus), the pressure due to the same liquid at two different temperatures, and use the relation (i) of Section 2.18. We describe here the experimental details of Dulong and Petit's method as improved successively by Regnault arid others.

The method uses the apparatus shown in Fig. 2.15. We have two vertical limbs/tubes AB and CD (about 1.5 metre) of a U-tube. The tubes are made of steel and are joined by narrow cross-tube AE and DG to wider, vertical glass tube EF and GH lying side by side at the top. The tubes AB and CD are connected together at the bottom by a narrow flexible iron tube BC. This enables the hot and the cold limbs of the U-tube to freely expand and contract quite independently of each other. It also leaves the tube AE and GD unaffected (*i.e.*, at the same level and horizontal).

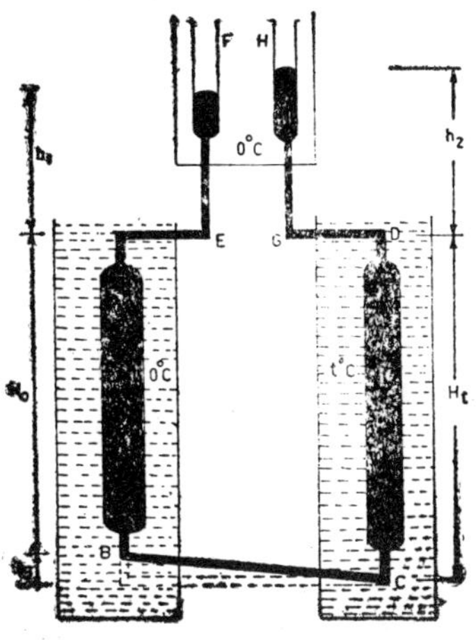

Fig. 2.15

The U-tube is filled with mercury up to about the middle of the tubes EF and GH. One of the limbs, AB, together with top wide tubes EF and GH are immersed in a bath of ice-cold water. The other limb CD, is enclosed in a bath of a suitable oil. On heating the oil bath, the mercury of the limb CD as well as the limb CD itself, expands whereas of the other tube-limb remains at lower temperature. This results in the portion C going down as compared to the end B and BC will therefore no longer be horizontal.

Maintaining the oil bath at a desired constant temperature t°C and allowing the mercury columns to attain steady heights h_1 and h_2 in the top tubes (see figure), the heights (i) H_0 the height of mercury column AB. (ii) h_1, the height of mercury in EF, (iii) h_2 the height of mercury in GH. (iv) H_t the height of mercury column in CD and (v) h_3, the difference in levels of B and C are carefully measured.

Now, since the open ends of U-tube F and H are exposed to the atmosphere we can write the pressure equation for the two portions as hydrostatic pressure due to mercury at B = hydrostatic pressure due to mercury at C.

or, $$h_1\delta_0 g + H_0\delta_0 g + h_3\delta g = h_2\delta_0 g + H_t\delta_t g$$

where δ_0, δ_t and δ are the densities of mercury at 0°C, t°C and mean temperature (room temperature, t^1) respectively. Then

$$h_1\delta_0 g + H_0\delta_0 g + \frac{\partial_0 h_3 g}{1+\gamma_m t^1} = h_2\delta_0 g + \frac{H_t\delta_0 g}{1+\gamma_m t}$$

or $$h_1 + H_0 + \frac{h_3}{1+\gamma_m t^1} = h_2 + \frac{H_t}{1+\gamma_m t}$$

whence γ_m can be obtained. Here we have substituted for $\delta_t = \delta_0(1+\delta_m t)$ and $\delta = \delta_0/(1+\gamma_m t^1)$.

The only snag in the improved apparatus being the value of t^1 but the term involving t^1 is quite Small and hence does not affect the result much. Regnault carried out experiments for determining γ_m for mercury over the range of temperature 0°C to 350°C and found that it was a function of temperature *i.e.*, $\gamma_m \equiv \gamma_m(t)$, given by

$$\gamma_m = 0.00017905 + 2.52 \times 10^{-8}t$$

The method described above was quite successful in removing drawbacks like:

(a) measurement of small difference in H_0 and H_1 as well as h_1 and h_2

(b) mixing up of hot and cold mercury through bottom of the tube.

(c) error due to tubes being not exactly horizontal, and

(d) balancing column at same temperature.

(ii) Modified Callendar and Moss Method : In view of the importance of mercury as a calibrating liquid Callendar and Moss realised that the greater the heights of the mercury columns in the cold and hot limbs the more correctly could the difference $H_t - H_0$ between the two be measured. Thus they decided to use much taller columns of mercury (but not in single vertical length which would have needed longer baths and affected the density of mercury due to compression) by using multiple columns (*i.e.*, number of pairs of cold and hot columns connected together in series as shown in Fig. 2.16). Like Regnault they also used tubes of steel of length 2 m each connecting each of the six pairs used, by narrow tubes of 1 mm in diameter.

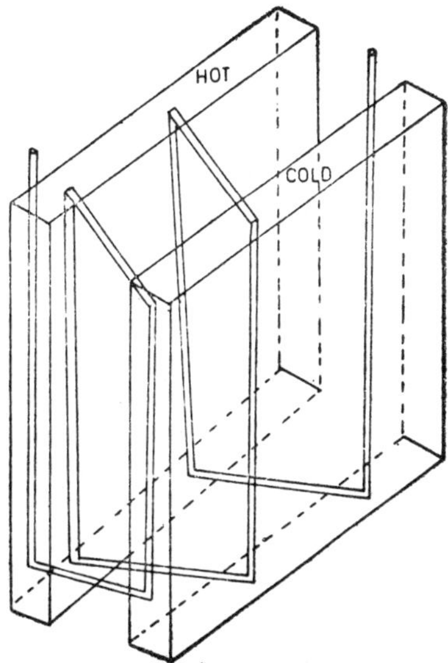

Fig. 2.16

Further, they used motor-driven stirrers in both the baths to ensure *uniformity* in their temperature which was measured by standardised platinum resistance thermometer with a large element extending over mercury columns. They also measured the difference in heights of the two columns on a standard invar scale (to avoid variations in the scale for room temperature fluctuations). They could thus obtain an accuracy of 1 m 10000 in the range of temperature 0° to 300°C. The value γ_m for mercury was found to be $(18056 + 1.24\ t) \times 10^{-6}$ and for 0 to 100°C it was 0.000182.

(iii) Other Methods : Coefficient of expansion of liquids, with sufficient accuracy, can be determined by the use of containers with exceptionally low coefficient of expansion such as silica or quartz. Or, a container with known cubical coefficient of expansion is filled with requisite amount of mercury whose expansion in volume is just the same as that of the container so that available volume of the vessel remains constant at all temperatures. Now this vessel will allow determination of real coefficient of expansion.

II. Determination of the Coefficient of Apparent Expansion of a Liquid

(i) *By volume thermometer or dilatometer* : A dilatometer essentially consists of a large bulb with a graduated stem as shown in Fig. 2.17 a. The liquid, whose coefficient of apparent expansion, γ_a is to be determined is filled in the dilatometer, which may then be placed in the bath maintained at a desired temperature, so that

$$\gamma_a = \frac{\text{observed increase in the voume of the liquid (i.e., liquid expelled}}{\text{initial volume of the liquid} \times \text{rise in temperature}}$$

This method is particularly suitable for volatile liquids of low specific gravity and in the study of changes in density of liquids with temperature.

(ii) *By weight thermometer or weight dilatometer:* This is the gravimetric method (that concerns with weight). Since weights can always be determined to a higher degree of accuracy than the volume, it is a better method based on the comparison of the densities of experimental liquid at two temperatures. The liquid is filled in the dilatometer (which is clean and dry) by dipping the capillary of the tube in the liquid. On heating the thermometer, air is pusher off and on cooling liquid enters into the bulb.

The method can be described by following observations :

(a) Weight of empty thermometer = m

(b) Weight of thermometer + liquid at $t_1{}^0$ C = M_1

(c) Weight of thermometer + liquid at $t_2{}^0$ C = M_2

$\therefore$ mass of liquid at $t_1{}^0C = M_1 - m = m_1$

and mass of liquid at $t_1{}^0C = M_2 - m = m_2$.

(d) Let the volume of the liquid at t_2°C and also at t_2°C = V

Then if the densities of liquid at $t_1{}^0$ and $t_2{}^0$C be δ_1 and δ_2 respectively

$$m_1 = V\delta_1 \text{ and } m_2 = \delta_2 V$$

or

$$\frac{m_1}{\delta_1} = \frac{m_2}{\delta_2}, \text{ or } \frac{m_1}{m_2} = \frac{\delta_1}{\delta_2}$$

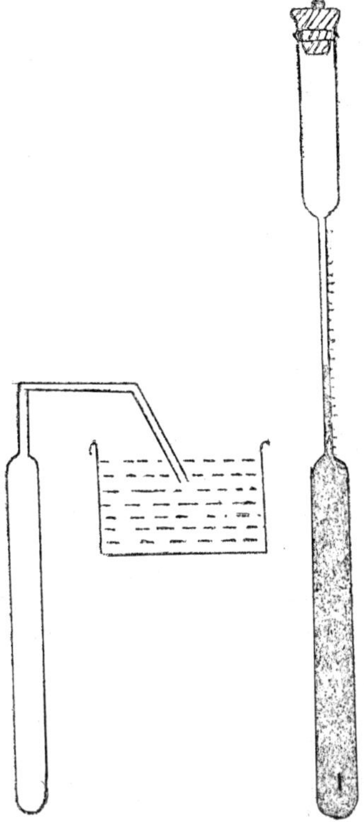

Fig. 2.17

But we know $\rho_1 = \dfrac{\rho_0}{1 + \gamma_a t_1}$

and $\rho_2 = \dfrac{\rho_0}{1 + \gamma_a t_2}$

So, $\dfrac{\rho_1}{\rho_2} = \dfrac{1 + \gamma_a t_2}{1 + \gamma_a t_1} \cong 1 + \gamma_a\ (t_a - t_1)$

$\therefore$ $\dfrac{m_1}{m_2} \cong 1 + \gamma_a (t_2 - t_1)$

or, $\gamma_a \dfrac{m_1 - m_2}{m_2\ (t_2 - t_2)}$

We may note : (i) weight of overflowing liquid can be determined directly using a catch pan for collecting it, (ii) it can be used at a series of rising temperatures, (iii) weighing can be done at room temperature.

(iii) By a Pyknometer : It is a density measurer and is an improved form of a weight thermometer shown in Fig 2.18. It allows easier operation of cleaning, filling, etc. It consists of a cylindrical glass bulb (6 cm long) to either end of which is fused a thermometer tube of a fine bore to form a U-tube. The free ends of this U-tube are further bent at right angles, at the top, in the opposite directions at the same horizontal level. One of the horizontal arm AC ends in a nozzle at C. A mark F is engraved on the other arm DE up to which the vessel will be filled in to have a known volume of the liquid.

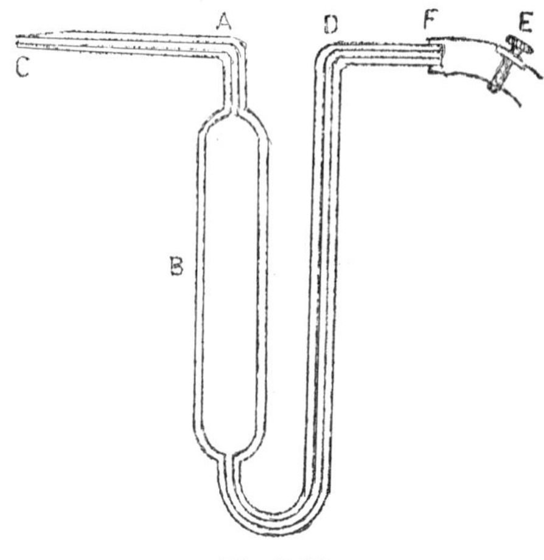

Fig. 2.18

To fill the Pyknometer a piece of rubber tubing with a pinch cock is attached at E with the nozzle dipped in the desired liquid. Air is sucked to fill in the apparatus with it.

The procedure for determining fa is similar to that of a dilatometer but by always maintaining liquid up to mark F.

(iv) By the Hydrostatic Balance or the Sinker Method : Mattiessen (1860) first employed this method based on the Archimedes principle. It consists of weighing a heavy body first in air and then in the liquid at two different temperatures.

Let the mass of the sinker in air = M

Mass of sinker in liquid at t_1°C = M_1

Mass of sinker in liquid at t_2°C = M_2

∴ loss of weight of sinker in liquid at t_1°C = $M_1 - M$

and, loss of weight of sinker in liquid at t_2°C = $M_2 - M$

Then, if be the volume of the liquid at both temperatures, then

$$\frac{\rho_1}{\rho_2} = \frac{M_1 - M}{M_2 - M} = \frac{\rho_0/(1 + \gamma_a t_1)}{\rho_0/(1 + \gamma_a t_2)} = \frac{1 + \gamma_a t_2}{1 + \gamma_a t_1}$$

$$\therefore \quad \frac{M_1 - M}{M_2 - M} \cong 1 + \gamma_a (t_2 - t_1)$$

or
$$\gamma_a = \frac{M_1 - M_2}{(M_2 - M)(t_2 - t_1)}$$

following the same procedure as detailed earlier in weight thermometer. The method has at least two .important drawbacks : (i) it is difficult to keep temperature constant while weighing, (ii) there is a surface tension pull on the wire where it meets the liquid surface.

2.20 EXPANSION OF WATER

In the case of water, there is a continuous variation in the slope of the volume-temperature graph indicating that the rate of change of volume with temperature varies continuously. But between 0° and 4°C it is found to contract (instead of expanding). Water has its minimum volume (*i.e.*, maximum density) at 4°C strictly speaking, 3.98°C). Its volume increases either side of this temperature. For this reason this temperature is also known as the temperature of inversion.

Curve in Fig. 2.19 shows relation between temperature and specific volume (volume per kg.) of water. We have dV/dT positive above 4°C and negative below 4°C.

No other pure liquid shows such an anomaly hence no satisfactory explanation has been put forward. A possible explanation for water is that there are three types of water molecules *viz.*, H_2O, $(H_2O)_2$ and $(H_2O)_3$, all having different specific volumes at different temperatures. Probably, at 4°C, the higher types preponderate, resulting in the maximum density of water at this temperature.

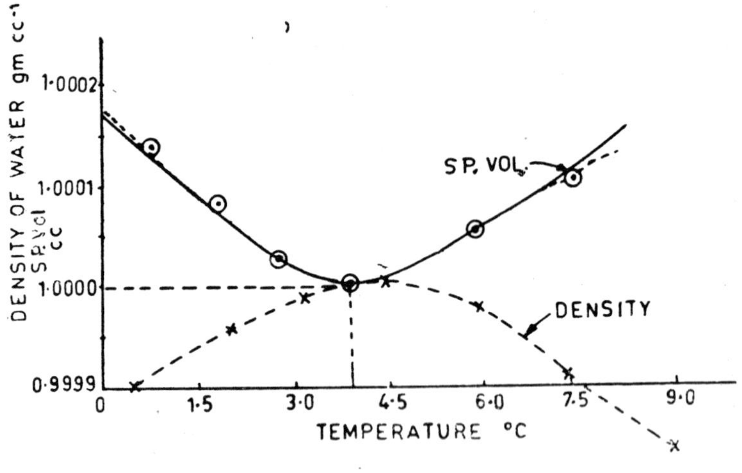

Fig. 2.19

2.21 HOPE'S EXPERIMENT

This experiment was designed (by Hope, 1805) to show that water possesses maximum density at 4°C.

It consists of a tall glass (or metal) cylinder C (Fig. 2.20) with two sensitive thermometers A and B fitted horizontally at the top and the bottom side of the cylinder. An annular trough, enclosing the cylinder at its mid portion, is filled with a freezing mixture while the cylinder itself is filled with water.

First, the water in the mid part of the cylinder cools by conduction, becomes denser and sinks down while the warmer water from below comes up. As a result, whereas the lower thermometer B goes on recording, first a rapid and then a gradual fall in temperature, the upper thermometer. A, shows practically steady temperature. This state of affair continues till the water below the mid-part of the cylinder falls to 4°C. It is shown by the graph in Fig. 2.21.

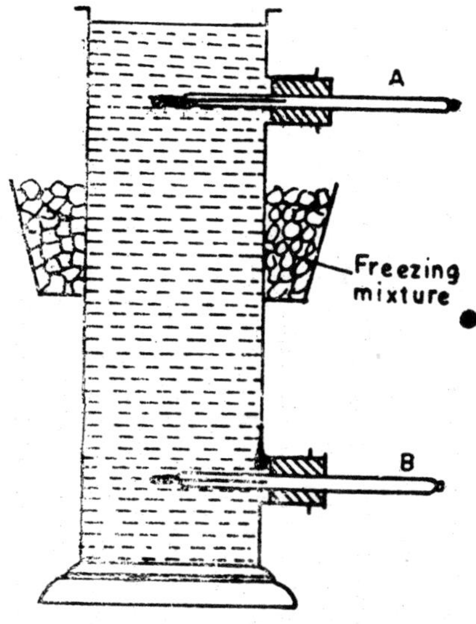

Fig. 2.20

After this, as the water in the region of the trough gets cooled to a temperature lower than 4°C, it becomes lighter than the water below and hence can no longer sink down. Thermometer B thus shows a constant temperature of 4°C whereas upper thermometer A continues to record a fall in temperature until it reads 0°C. It is thus clear that water has its maximum density at 4°C, temperature at which the two, curves meet and cross each other.

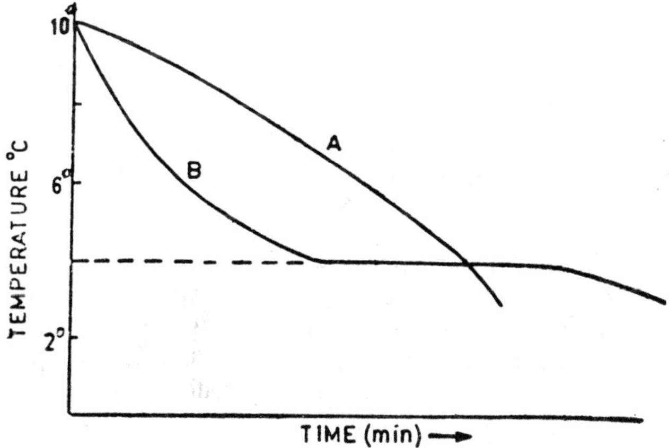

Fig. 2.21

Eventually, due to loss of heat by conduction from both parts, the whole of the water in the cylinder gets frozen into ice from the top downwards.

2.22 JOULE AND PLAYFAIR'S EXPERIMENT

This is a more accurate method than Hope's for the determination of the temperature of maximum density of water.

The apparatus is based on the principle of balancing columns. The apparatus (Fig. 2.22) is thermally insulated from the surroundings. Cylinders A and B are filled with water (trough F and tap T closed) at a temperature little below the inversion temperature and then some water is added in to cylinder B so as to raise the temperature of the water in it to about as much above the maximum density temperature as temperature in A is lower than it. The water in both the cylinders is kept well stirred to ensure uniformity of temperature.

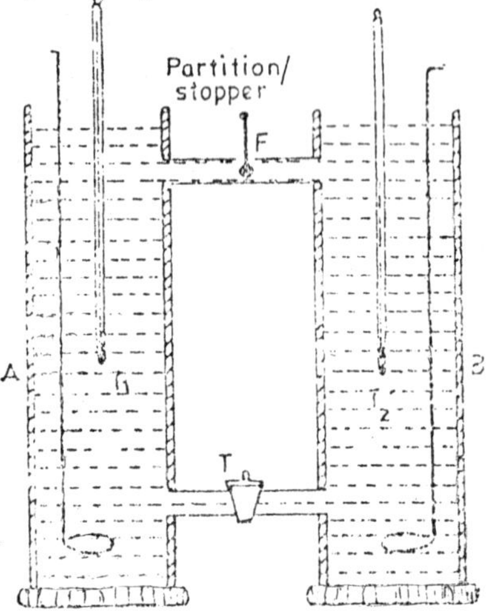

Fig. 2.22

The tap is now opened and the partition in trough F is removed to put the two water columns in communication and the direction of the flow of water in the trough is detected by float at F (of glass—bulb floating on the water surface). This direction will be opposite to the direction of flow through the pipe at the bottom, where it will always be from a higher to a lower density.

For example, if flow of water is from A to B at the bottom pipe (indicating that density of A, the cold water is higher than that of the density of B, the hot water) then we have the inequality

$$T_1 - T_1 > T_2 - T_t$$

where T, is the maximum density temperature. So that, taking the density temperature curves of water to be symmetrical about the temperature of maximum density we have

$$2T_t > T_1 + T_2$$

or, $$T_t > \frac{T_1 + T_2}{2}$$

but, if flow of water is from B to A at the bottom pipe the condition becomes

$$T_i < T_1 < T_2 - T_t$$

or, $$T_i < \frac{T_1 + T_2}{2}$$

In the ideal case, when the float moves neither way there is no flow of water between the two cylinders, indicating that the density of water in both is the same, and

$$T_t - T_1 = T_2 - T_t$$

or $$T_t = \frac{T_1 + T_2}{2}$$

i.e., the mean of the two temperatures would thus be the correct maximum density temperature of water. Since in practice, it is almost impossible to obtain such an ideal state, the rate of travel of the float is measured for different mean temperatures in A and B and a graph is plotted between the two. The mean temperature, corresponding to zero rate of travel of the float gives the correct temperature at which water has its maximum density. We may note the following points also in connection with this experiment:

1. The temperature of maximum density decreases with increase lot pressure (it is 2°C at a pressure of 100 atmosphere, *i.e.*, a change of 0.025°C per atmospheric pressure change). The volume of 1 kg water at 4°C under one atmosphere is 1000'.028 cc.
2. The pressure of dissolved substances also tends to lower the maximum density temperature of water.

A consequence of the anomalous expansion, of water-Freezing of water in ponds and lakes. A direct consequence of the maximum density temperature of water is that of water in a pond. Here surface layers start cooling due to cold air above. As the upper layer of water cools it becomes heavier hence goes down and the process continues till the temperature of the bottom layer falls up to 4°C. When surface layer cools further its temperature falls to zero and ultimately freezes without going down (as this is lighter now) whence aquatic life survives.

2.23 OTHER CASES OF ANOMALOUS EXPANSION

There are quite a few aqueous solutions that also exhibit a contraction in volume (instead of expansion) in the range of temperature 0° to 4°C. There are other substances also like antimony, resolidified silver iodide, bismuth and iron which, like water, expand when they solidify. Thus when we plot a graph between volume of a certain mass of a substance at different temperatures we find the graph, which is, in many cases, not linear. Thus it is not correct to write

$$V_t = V_0 (1 + \alpha t)$$

But, should be

$$V_t = V_0 (1 + \alpha t + bt^2 + ct^3 + ...)$$

and is really important when we deal with the liquids which do not expand regularly.

2.24 UNITS OF MASS AND VOLUME

The founders of the metric system defined the unit of roast, *i.e.*, a kilogram, as the mass of one litre of pure water at the temperature of maximum density. The standard platinum kilogramme was thus made by Borda and preserved at the International Bureau of Metric Weights and Measures, Sevres (Paris). But, now this platinum standard exceeds the water standard by 28 milligram. So, the volume of 1 kg of water at its maximum density now, is 1000.28 cc. Hence to corroborate the definition of mass, a new definition to volume is given (one litre is equal to volume of 1 kg water at its maximum density). The new litre (= 1000 ml) equals 1.000028 conventional litre (1000.028 cc). Thus, strictly speaking 1 ml $\neq$ 1 cc.

2.25 PRACTICAL APPLICATIONS OF LIQUID EXPANSION

I. Cases where Expansion is Helpful

Among the useful applications of thermal expansion of liquids are

(i) Liquid in glass thermometers

(ii) Mercury compensated pendulum

(iii) Liquid thermostats and

(iv) The central heating system.

We have discussed first two applications earlier. Here we shall deal with the latter two.

Liquid Thermostats : As indicated earlier, it is a device that regulates the rate of supply of heat to a liquid bath, furnace, laboratory device etc.

Liquid thermostats are generally used up to 100°C and are more delicate in control of temperature ± 1°C to 2°C. (It is not possible to have absolute constancy as the device comes into operation only when the desired temperature has just exceeded, and, similarly restores it back automatically when the temperature falls just below the desired value.

A simple *Toluene thermostat* is shown in Fig. 2.23 and is used to regulate the temperature of a gas heated bath. It consist of a large bulb (or a spiral) B^1 which is given a shape such that it presents the maximum surface area to the liquid of the bath in which it is placed. The bulb is filled with a liquid (toluene) having a high boiling point and a high coefficient of expansion. The bottom of the bulb is connected to a long verticle stem F which ends at the top into another cylindrical bulb C. A tube AJ bent into right angle is fitted into C and ends in a jet J. Another small vertical tube BD, provided with a stopcock T, connects AJ &ad a side tube ED from the bulb C.

The gas enters at A and emerges out of the Jet at J and then passes through the side upto DE into the burners heating the bath. When the toluene of the bulb B' expands due to rising temperature of the bath the mercury in the stem is forced up into C and closes the jet J thus, blocking the passage of the gas, so that gas can reach the burner, but only through T, a very depleted supply. The bath temperature thus falls, whence toluene contracts allowing the jet to open and increase the burner feed.

Now as is clear, the temperature at which jet should close depends on the level of jet J which is made adjustable in position for a desired value of constant temperature.

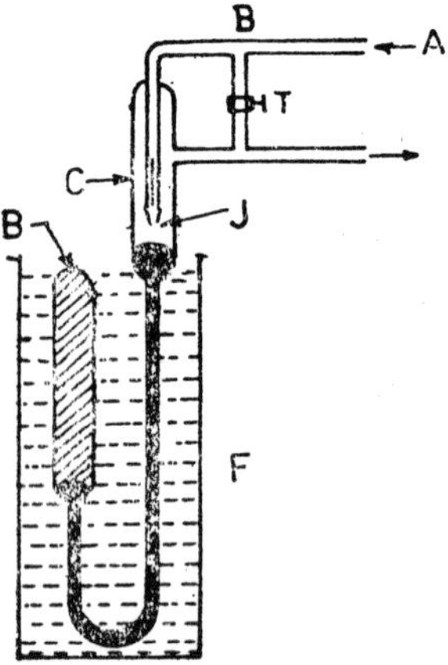

Fig. 2·23.

Fig. 2.23

II. Cases where Expansion has to be Corrected for

Among these are (i) emergent column in a mercury thermometer (ii) reduction of barometric height to 0°C, etc. Here we will consider the accuracy we can have for a barometric height.

The brass scale of the barometer is usually calibrated at 0°C. So that readings on it are correct only when the scale and the mercury in the tube both are at 0°C. But we read the barometer at the room temperature t°C when both the brass scale and the mercury are in the expanded state. Therefore we need to apply corrections for:

(i) Expansion of the brass scale—Let the barometric height at t°C be H_t and the true barometric height be H and t t°C then as shown in section 2.14

$$H_t = H(1 + \alpha t)$$

(ii) Correction for expansion of mercury—I$_f$ ρ_t be the density of mercury at t°C then pressure of the atmosphere = pressure due to H, height of the mercury column.

$$= H_t \cdot \rho_t \cdot \gamma$$

$$= H(1 + \alpha t)\, \rho_t \gamma$$

Now, if H_0 be the true barometric height at 0°C corresponding to height H_1 and if ρ_0 be the density of mercury at 0°C, we have

Pressure of the atmosphere = $H_0 \rho_0 \gamma$

i.e., $H_0\rho_0\gamma = H(1 + \alpha t)\rho_t\gamma$

$$\therefore \quad H_0 = H(1 + \alpha t)\frac{\rho_t}{\rho_0}$$

$$= H\,\frac{(1 + \alpha t)}{(1 + \gamma t)}$$

$$\therefore \quad H_0 \cong H(1 + \alpha t)\,(1 + \gamma t)$$

$$H_0 \cong H[1 + (\alpha - \gamma)\,t]$$

where $\dfrac{\rho_t}{\rho_0} = \dfrac{1}{(1 + \gamma t)}$, γ being coefficient of absolute expansion of mercury.

Now, since $\gamma > \alpha$

$$H_0 \cong H[1 - (\gamma - \alpha)t]$$

for mercury $\gamma = 0.000182$ and α for brass = 0.000019

$$\therefore \quad H_0 = H(1 - 0.00016\ t)$$

is the true barometric height at 0°C.

SECTION III: EXPANSION OF GASES

2.26 VOLUME AND PRESSURE COEFFICIENTS OF A GAS

Gases too have no shape of their own and their properties are also tame in all directions. But, whereas liquids have a definite volume and have a clearly discernible surface of separation from the surrounding medium, gases have no definite volume of their own nor do they have any free surface as such. Further, their expansion due to heat is enormous in comparison to that of solids and liquids and they are highly expansible/ compressible for changes in pressure.

Thus in the case of gases : (i) We have to contain them in completely closed vessels, (ii) We have to deal with three variables, viz., temperature, pressure and volume. This naturally complicates the thermal expansion of gases. Yet, a study of it is by far more important as (i) it leads to formulation of all universal gas laws, and (ii) it furnishes us with a most accurate, reliable and easily reducible scale of temperature.

Of the three variants we can take two at a time keeping the third constant and have three different relations for a gas.

(i) *Volume-temperature relation at constant pressure* (The Charles law). When a given mass of a gas is heated at a constant pressure its volume increases proportionately. The fractional increase in volume of a given mass of gas per degree (celsius) rise of temperature at constant pressure is the volume coefficient of a gas, γ_v.

i.e., $$\gamma_v = \frac{V_t - V_0}{V_0 t} \cong \frac{dV}{V.t}$$

or $$V_t = V_0 (1 + \gamma_v t).$$

(ii) *Pressure-temperature relation at constant volume* (also known as Charles constant volume law). When a given mass of gas is heated, maintaining its volume constant, its pressure increases proportionately with its temperature, and the fractional increase in pressure of a fixed mass of a gas at constant volume is the pressure coefficient of a gas, γ_p.

i.e., $$\gamma_p = \frac{P_t - P_0}{P_0 t} \cong \frac{dP}{P.t}$$

or $$P_t = P_0 (1 + \gamma_p t).$$

(iii) *Volume-pressure relation at constant temperature* (Boyle's Law). When a pressure of a given mass of gas changes, its volume changes inversely. But this relation is not needed in our present scheme of study.

Above relations for the two coefficients may have subscripts indicating value of the parameters at t°C and 0°C, for convenience.

2.27 DETERMINATION OF VOLUME COEFFICIENT OF A GAS

(i) Gay Lussac's Method : Gay Lussac was first to determine the volume coefficient of a gas. His apparatus (shown in Fig. 2.24) had a glass bulb with a graduated long stem. The whole bulb was first filled with mercury and fitted up side down, with its stem enclosed in a wider tube and packed all over with calcium chloride. The mercury was then gradually shaken out by means of a platinum wire inserted through the stem leaving just a small thread in the stem, by way of an index. (The process ensures filling the bulb with dry air). It was then placed in a bath of melting ice with its stem perfectly horizontal. When the pressure of the enclosed air became equal to that of the atmosphere the volume V_0 was noted by the index point. Next bath was heated to t°C, so mercury index was pushed out and attained a steady position when the pressure inside the bulb again became one atmosphere. Thus, volume V, is read and y, is calculated (for air, $\gamma_u = 1/268 \cong 0.00375$). The error in the value of γ_v (accurate value being $1/273.16 \cong 0.00366$) may be due to (i) a thin film of water remained stuck inside the bulb which may become vapour on heating (ii) the mercury index may not completely seal communication of inside ail from outside (iii) the mercury index needs a small pressure difference for its movement.

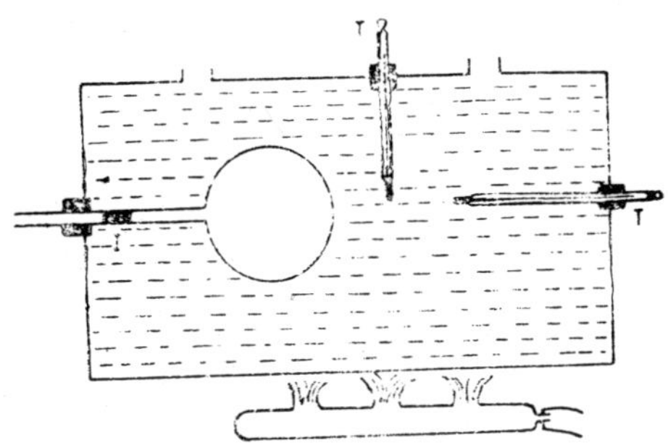

Fig. 2.24

(ii) Regnault's Method : It consisted of a large bulb A (Fig. 2.25) surrounded by a jacket and connected to a mercury manometer BCDE immersed in a constant temperature bath W. The limb BC was accurately calibrated to measure the volume of the bulb and connecting tube. To avoid surface tension effects on the readings the limb DE had same bore as that of the limb BC. A tap F is provided to let out the mercury when desired.

Bulb A was filled in with dry air through G (First evacuated severally and then filled with air) and allowed to stand at 0°C. The tube G ensures that the pressure of the air is one atmosphere with mercury in the two limbs standing at the same level (by allowing air in through G or running out some mercury through F). The volume of the gas is then recorded as V_1.

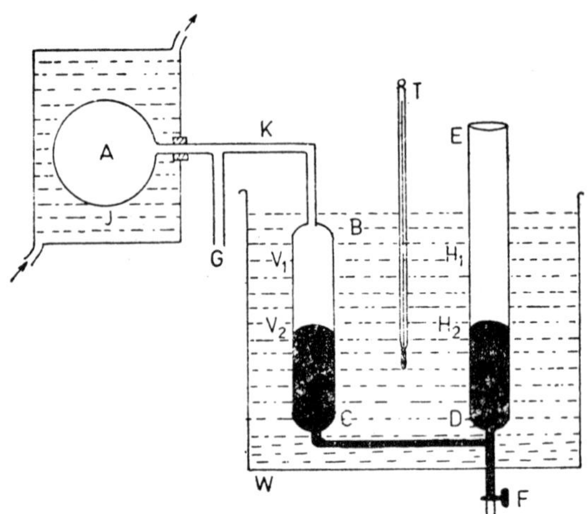

Fig. 2.25

The Chamber J is then heated by circulating steam till gas in A acquired temperature of t°C which upsets mercury levels in the limbs. Pressure is again set equal to one atmosphere and volume of the gas is read off. Still it was found that the accuracy of the apparatus goes down as the bulb temperature is increased.

Regnault attempted suitable corrections for various sources of errors like (i) different temperatures of the gas in bulb and in connecting tube etc., (ii) expansion of bulb A etc. As a result, he succeeded in showing quite clearly that expansion of no gas was perfectly uniform and that except perhaps hydrogen the volume coefficient of no gas was exactly 1/273.16 in the range 0° to 100°C of temperature.

He further, found that the volume coefficient increased with the pressure though the increase was small. His results for different gases in the range 0–100°C for $\gamma_,$ are:

Air	0.003671	H_2	0.003661
CO_2	0.003700	H_2O	0.003719
CO	0.003669	SO_2	0.003903.

(iii) Jolly's Modification of Regnault's Method : In this modified method. Jolly arranged the bulb A vertically, which was connected to a graduated manometer by a narrow horizontal tube K as shown in Fig. 2.26. DE limb was free to move vertically as it was connected by a flexible rubber tubing. A three-way tap S was used to facilitate communication between K or G to BC. He also noted barometric pressure intermittently. He was thus able to do away with:

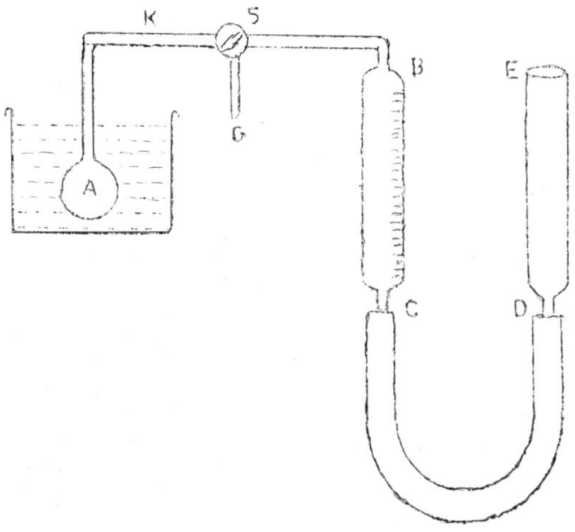

Fig. 2.26

(i) the problem of placing bulb A in the jacket,

(ii) quantity of gas in the connecting tube being at different temperature,

(iii) problem of adjusting mercury levels,

(iv) necessity of surrounding bulb A in the melting ice bath, initially.

Further communication tap S helped (though it may cause leakage problems) to have desired communication.

Let the bulb A be kept in the bath at t_1°C and then at t_2°C. If the corresponding volumes be V_1 and V_2 respectively, then

$$V_1 = V_0 (1 + \gamma_v t_1)$$

$$V_2 = V_0 (1 + \gamma_v t_1)$$

$$\therefore \qquad \gamma_v = \frac{V_2 - V_1}{V_1 t_2 - V_2 t_1}$$

γ, can also be obtained if a curve (in this case a straight line) is obtained between volume and temperature. The intercept of this line gives V_0 at the volume axis and when extrapolated to zero of volume, we get

$$0 = V_0(1 + \gamma_v t_1)$$

or
$$t = -\frac{1}{\gamma_v} = -273.16°C$$

or the zero on the absolute scale of temperature (also see section 1.12 for Callendar's, constant pressure air thermometer, for the improved version for measurement of γ_v).

2.28 DETERMINATION OF PRESSURE COEFFICIENT OF A GAS

(i) Regnault's Experiment : Regnault used the same apparatus-as shown in Fig. 2.25. with a slight modification in the design of the manometer. His values for γ_p were

air	0.003667	N_2	0.003676
CO_2	0.003725	N_2O	0.003676
CO	0.003675	SO_2	0.003845
H_2	0.0036678.		

Following were his additional observations:

(a) no gas had exact value of 1/273.16 for γ_p,

(b) value of γ_p remained same at other ranges of temperatures also,

(c) value of γ_p increased at higher pressures, *i.e.*, it was not same in different pressure ranges.

(ii) Laboratory Method, Jolly's Apparatus : This apparatus (see Fig. 2.27) was different from the earlier in that it had a mark M on the limb BC, Further, the reservoir R had a side tube E of the same bore as BC. The screen S' was also placed between A and BC to avoid any heating of the mercury columns.

Now as the pressure varies it is read as the difference in height of mercury h^2 in the two limbs BC and E.

Thus, $P_0 = H + h_0$

and $P_t = H + h_t$

where H is the barometric height whence

$$\gamma_p = \frac{P_t - P_0}{P_0 t}$$

Alternatively, if we do not start the experiment by measuring pressure at 0°C we can plot pressure-temperature relation to obtain a straight line AB as shown in Fig. 2.28. The equation of this line is

$$P = P_0\gamma_p t + P_0$$

(and $V = V_0\gamma_v t + V_0$ if P is replaced by V for section 2.27).

The line cuts P axis at P_0 and when extrapolated to cut temperature axis we have OD when pressure of the gas reduces to zero. This temperature on the celsius scale is, again –273.16°C and

$$Y = 1/273.16$$

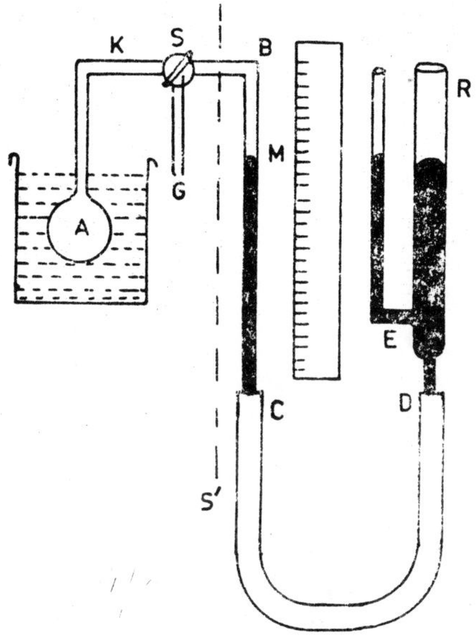

Fig. 2.27

Drawbacks of the method being : (i) the expansion of bulb A, if not taken care of, (ii) gas in the connecting tube being at different temperature (this dead space effect is not serious in volume-temperature relation), (iii) variation in atmospheric pressure affects the result.

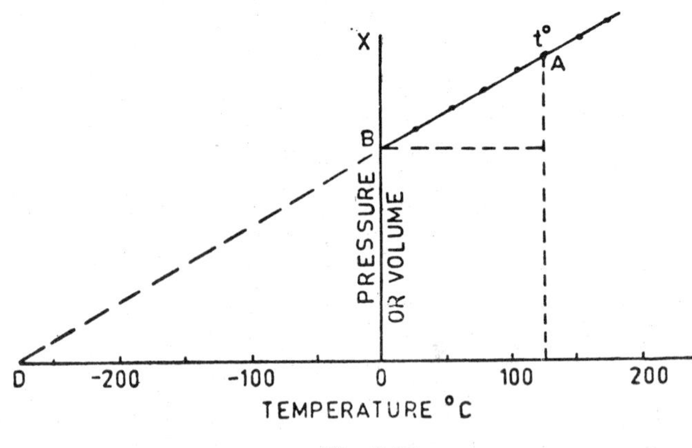

Fig. 2.28

2.29 EQUALITY OF VOLUME AND PRESSURE COEFFICIENTS

Suppose we start with a volume V_0 at 0°C and pressure P_0 for a given mass of a gas. Now, hear it at constant pressure up to t°C, so that the new volume is $V_t = V_0(1 + \gamma_v t)$. Now, keeping the temperature of the gas at t°C let it be compressed back to its original volume V_0. In the process, let the pressure becomes P_t, then

$$P_0 V_t = P_t V_0. \qquad \text{...(i)}$$

This new pressure could be arrived by heating the gas up to t°C at constant volume V_0, *i.e.*,

$$P_t = P_0(1 + \gamma_p t). \qquad \text{...(ii)}$$

Substituting values of V_t and P_t in the relation (i) we have

$$P_0 V_0(1 + \gamma_v t) = P_0 V_0(1 + \gamma_p t) \quad \text{or } \gamma_v = \gamma_p$$

Thus for a Perfect gas, the volume coefficient is equal to pressure coefficient. The value given by Heuse and Otto is 1/273.16 or 0.003660 (degree celsius)$^{-1}$.

2.30 ABSOLUTE ZERO AND ABSOLUTE SCALE OF TEMPERATURE

Absolute zero is the temperature at which pressure of the gas at constant volume is reduced to zero. This temperature, –273.16°C, is absolute zero, as presumably is the lowest temperature attainable.

This conception of absolute zero has indeed proved to be a most useful one, as we shall see, as we proceed along further, but all gases actually liquefy, or even solidify, before they can be cooled to this temperature. This does not, however, invalidate the relationship between volume and temperature or between pressure and temperature at higher temperatures.

The scale of temperature with –273°C (strictly speaking –273.16°C) as absolute zero and similar fundamental interval as in Celsius scale we have what is known as gas scale or the absolute scale of temperature.

From Fig. 2.28 we have

$$DE = DO + OE = \frac{1}{\gamma} + t = 273.16 + t$$

so that $V \text{ or } P \propto (273.16 + t)$

$$\propto T \qquad ...(iii)$$

where T = 273.16 + t. This relation can be stated in words as:

"The volume of a given mass of a gas, under constant pressure, is directly proportional to its absolute temperature", and

"The pressure of a given mass of a gas at constant volume, is directly proportional to its absolute temperature."

These are *just* the enunciations of the two Charles laws. From relation (iii)

$$V_1/V_2 = T_1/T_2$$

and $P_1/P_2 = T_1/T_2$

2.31 APPLICATION OF MANSION OF GASES

The foremost application of the thermal expansion of gases is the gas thermometer. The equality of γ_p and γ_v his led us to the formulation of the all important Charles' laws which have given is the conception of absolute scale of temperature, the readings on which are identical with those JOB Kelvin's thermodynamic scale of temperature. The two laws have also led to the establishment of an equation of state involving a universal gas constant.

Another useful application of gas-expansion is the liquefaction of gases under high pressure, which has led to the production of low temperatures.

The conception of perfect gas has enabled-us to calculate the efficiency of a Carnot reversible engine also.

SOLVED EXAMPLES

Example 1:

What do you understand by mean coefficients of linear expansion? What error do you expect if you take the expansion coefficient to be the same?

Solution:

If the expansion is stated as a fraction of h at a lower temperature t_1 of the temperature range t_1 to $t_2°$, the mean coefficient over this range is defined as

$$\alpha_m = \frac{1}{l_1}\frac{(l_2 - l_1)}{(t_2 - t_1)}$$

Now, consider ε_0 be the linear coefficient of expansion for the range 0° to t_1°C α_1 for $t_1°$ to t_2°C, α_2 for $t_2°$ to t_3°C etc. and that the ranges are small enough so that we may take $\alpha_n \cong \alpha_{n-1}$. If .similar notations are used for lengths also we have

$$l_1 = l_0(11 + \alpha_0 t)$$

$$l_2 = l_3[1 + \alpha_1(t_2 - t_1)]$$

$$l_3 = l_2[1 + \alpha_2(t_3 - t_2)]$$

$$l_n = l_{n-1}[1 + \alpha_{n-1}(t_n - t_{n-1})]$$

or $$l_1 = l_0 = \delta l_6 = l_0\alpha_0 t_1$$

$$l_2 = l_1 = \delta l_1 = l_1\alpha_1(t_1 - t_2) \qquad \text{...(i)}$$

$$l_n - l_{n-1} = \delta l_{n-1} = l_{n-1}\alpha_{n-1}(t_n - t_{n-1})$$

adding the set of equations (i) we have

$$l_n - l_0 = \delta l_0 + \delta l_1 + \ldots \delta l_{n-1} = l_0\alpha_0 t_1 + l_1\alpha_1(t_1 - t_1) + \ldots$$

if we take, for the sake of simplicity,

$$t_n - t_{n-1} = t_{n-1} - t_{n-2} = \ldots = \delta t;$$

$$l_n - l_0 = l_0\alpha_0\delta t + l_1\alpha_1\delta t + \ldots + l_{n-1}\alpha_{n-1}\delta t$$

or $$\frac{l_n - l_0}{n} = \delta t \left[\frac{l_0\alpha_0 + \alpha_1 l_0 + \ldots \alpha_{n-1} l_{n-1}}{n}\right]$$

$$\therefore \quad \left.\begin{matrix}\text{mean}\\ \text{expansion}\end{matrix}\right\} = \sum_{i=0}^{n-1} \frac{\alpha_t l_t}{n} \delta t$$

Now, applying second approximation *viz.*, $\alpha_0 = \alpha_1 = \alpha_2 \ldots \alpha_{n-1}$ we have mean expansion due to rise in temperature δt

$$= \alpha\delta t \left[\frac{l_0 + l_1 + l_2 + \ldots + l_{n-1}}{n}\right]$$

As $l_n > l_{n-1} > l_{n-2} > \ldots > l_2 > l_1 > l_0$, the mean expansion will have larger value if we take more and more of division.

The same can be shown also by

$$l_1 = l_0 (1 + \alpha_0 t_1)$$

and $$l_2 = l_1 [1 + \alpha_1(t_2 - t_1)]$$

or $$l_2 = l_0 (1 - \alpha_0 t_1) [1 + \alpha_1(t_2 - t_1) + \alpha_0 t_1)$$

$$= l_0 [1 - \alpha_1 t(t_2 - t_1) + \alpha_0\alpha_1 t_1(t_2 - t_1) + \alpha_0 t_1]$$

$$\cong l_0 [1 - \alpha_0 t_2 + \alpha_0^2 t_1(t_2 - t_1) \; [\text{for } \alpha_1 = \alpha_0]$$

i.e., $l_2 - l_0$ has additional term $\alpha_0^2 t_1 (t_2 - t_1)$ though small.

It can be seen in yet another way

$$\alpha = \frac{1}{l}\frac{dl}{dt}$$

or $$\alpha dt = \frac{dl}{l}$$

$$\therefore \quad \int_{t_1}^{t_2} \alpha dt = \int_{t_1}^{t_2} \frac{dl}{l}$$

or $$\alpha(t_2 - t_1) = \log l_{t2}/lt_1$$

or $$l_{t2} = l_{t1}\, e^{\alpha(t_2 - t_1)}$$

$$= l_t \left[1 + a(t_2 - t_1) + \frac{\alpha^2 (t_2 - t_1)^2}{2!} + \ldots\right]$$

Thus, successive terms being larger, you add the error as you increase the range.

Example 2:

What corrections you would like to effect for measuring real expansion coefficient with a dilatometer?

Solution:

Let the volume of the dilatometer bulb and one division be V and v respectively and let subscripts to them indicating the temperature at which the quantity is being recorded and let c and γ be volume expansion coefficients of dilatometer material and the liquid. Then

Volume occupied by the liquid at 0°C is $V_0 + n_0v_0$ and at t°C is $V_t + n_tv_t$

$$\text{but} \qquad V_t = V_0(1 + ct) \text{ and } V_t = v_0(1 + \gamma_t)$$

$$\therefore \qquad V_t + n_tv_t = V_0(1 + ct) + n_tv_0(1 + ct)$$

$$\text{and also,} \qquad = (V_0 + n_0v_0)(1 + \gamma t)$$

$$\text{or} \qquad (V_0 + n_tv_0)(1 + ct) = (V_0 + n_0v_0)(1 + \gamma t)$$

$$\text{or} \qquad \gamma = \frac{(n - n_0)v_0}{V_0 + nv_1)t} + c\left(\frac{V_0 + nv_0}{V_0 + n_0v_0}\right)$$

where n is the number of divisions up to which the liquid stands in the dilatometer stern.

The above expansion gives γ_{mean} in the range of temperature 0° to t°C. The true coefficient of expansion is

$$\gamma = \frac{1}{V}\frac{dV}{dt} = \frac{1}{V}\left\{c(V_0 + nv_0) + (1 + ct)v_0\frac{dn}{dt}\right\}$$

$$= \frac{c}{1 + ct} + \frac{v_0}{V_0 + nv_0}\frac{dn}{dt}$$

dn/dt can be obtained from a graph of n against and the coefficient is evaluated for each temperature t.

Example 3:

In a constant volume thermometer v_0 volume of the gas is exposed to a temperature 0°C. What corrections must be applied to the pressure?

Solution:

Let t°C be the temperature of the bulb and V be the total volume of the bulb and connecting tube. It is the gas in the connecting tube which is exposed to 0°C.

Now, let the temperature of the connecting tube be changed from 0° to t° without altering the conditions of the bulb and keeping the pressure constant, then

$$V\theta = V_0(1 + \gamma\theta)$$

and $$V_t = V_0(1 + \gamma t)$$

$\therefore$ $$v_t = v\theta\left(\frac{1 + \gamma t}{1 + \gamma\theta}\right)$$

Hence total volume of the gas at t°C is

$$V - v_0 + v\theta\left(\frac{1 + \gamma t}{1 + \gamma\theta}\right)$$

for t > θ, it is

$$V + \left[\frac{\gamma(t - \theta)}{1 + \gamma\theta}\right] v_0$$

Now, we keep the temperature constant and vary the pressure until the gas once again occupies the original volume V then, if P is the pressure observed in the instrument and P' be the corrected pressure which would have been observed if the connected tube were also at the same temperature as the bulb, we have

$$P'V = P\left\{V + \left[\frac{\gamma\,(t - \theta)}{1 + \gamma\theta}\right] v_0\right\}$$

or $$P^1 = P\left\{1 + \left[\frac{\gamma\,(t - \theta)}{1 + \gamma\theta}\right]\frac{v_0}{V}\right\}$$

Thus, $\frac{P}{V}\left[\frac{\gamma\,(t - \theta)}{1 + \gamma\theta}\right] v_0$ is the factor to be added to the observed pressure for corrected readings.

Example 4:

Two equal straight strips of two different metals are fastened together, parallel to each other, a small distance apart to form a bimetallic strip, show that it gets bent, both when heated and cooled considerably. Obtain the expression for its radius of curvature, when thus bent.

Solution:

Let 1 and 2 be two straight and equal strips of two different metals fastened parallel to each other at a distance d at 0°C as shown in the Fig. 2.29.

Then, on being heated to t°C, the two strips expand differently due to their different coefficient of linear expansion. So that bimetallic strip, as a whole gets bent into an are, of mean radius of curvature γ with the more expansible strip on the outer and the less expansible strip on the inner side.

If l_0 be their initial length at 0°C and α_1 and α_2 be the respectively coefficients of expansions then

$$l_1 = l_0(1 + \alpha_1 t), \;\; l_2 = l_0(1 + \alpha_2 t)$$

if φ be the common angle subtended by the bent strips then

$$l_2 = \varphi(\gamma_1 + d), \; l_1 = \varphi\,(\gamma_1)$$

or $$l_0(1 + \alpha_2 t) = \varphi\gamma_1 + \varphi d = l_1 + \varphi d$$

or $$l_0(1 + \alpha_2 t) = l_0(1 + \alpha_1 t) + \varphi d$$

or $$\varphi = \frac{l_0 t\,(\alpha_2 - \alpha_0)}{d}$$

but $$l_2 = \varphi\gamma_2 \text{ and } l_1 = \varphi\gamma_1$$

or $$\frac{(l_1 + l_2)}{2\varphi} = \frac{r_1 + r_2}{2} \cong \gamma$$

$\therefore$ $$r = \frac{l_1 + l_2}{2\varphi} = \frac{l_0\,(1 + \alpha_1 t) + l_0(1 + \alpha_2 t)}{2\varphi}$$

or, $$r = \frac{2l_0}{2\varphi} + \frac{l_0 t(\alpha_1 + \alpha_2)}{2\varphi}$$

$$\cong \frac{l_0}{\varphi}$$

$\therefore$ $$r = d/t\,(\alpha_2 - \alpha_1)$$

when this bimetallic strip is cooled, it bents, but now in the opposite direction.

Example 5:

A brass scale which is correct at 0°C gives the length of a steel rod as 41.628 cm when both rod and the scale are at 20°C. Calculate the true length of the rod at 20° C and at 0°C as measured by the scale if the latter remained at 20°C, coefficients of linear expansion of brass and steel being 0.000011°C ^{1} and 0.0000°C ^{1} repetitively.

Solution:

The brass length being correct at 0°C hence the reading 41.628 cm would really be correct at 0°C, which is to equal at 20°C to

$$l_{20} = 41.628\ (1 + 0.000019 \times 20)$$
$$= 48.644 \text{ cm}$$

Now if l_0 were the length of the steel rod at 0°C we have

$$l_{20} = l_0(1 + 0.000011 \times 20)$$

or $$41.664 = l_0(1 + 0.00022)$$

or $$l_0 = 41.635 \text{ cm.}$$

This would be the length of the rod, read on the brass scale as if it were at 0°C. Since however, it is at 20°C, the length 41.635 cm on it will have expanded by 41.635 × 20 × 0.000019 = 0.0158. So that, the scale division showing 41.635 cm will have moved 0.0158 cm towards the right. Hence the length of the steel rod at 0°C as shown by the scale at 20°C will be 42.635–0.158 = 42.619 cm.

Example 6:

The diagram shows an iron wire AB stretched inside a rigid brass framework and rigidly attached to it at both ends A B. The length of AB = 3 m at 0°C and diameter of the wire is 0.6 mm. What extra tension will develop in the stretched wire when the temperature of the system is raised to 40°C. Given $\alpha_{Fe} = 0.000012°C^{1}$, $\alpha_{brass} = 0.000018°C^{-1}$ *and* $Y_{Fe} = 2.1 \times 10^{11}$ *Newtons per square meter.*

Solution:

Clearly $l = 3\text{m}$

$$\therefore \quad l_{40} = 3(1 + \alpha_{Fe}40)$$

$= 3.00144$ m

and length of the frame

$$l_{40} = 3(1 + \alpha_{brass40})$$

$$= 3(1 + 72 \times 10^{-5})$$

$\therefore$ due to pull of the frame the extra increase in length

$$\delta l = 3(1 + 72 \times 10^{-5}) - 3(1 + 48 \times 10^{-5})$$

$$= 0.00072 \text{ m}$$

$\therefore$ Extra tension $F = \gamma a \cdot \dfrac{\delta l}{t}$

$= 0.142$ Newton.

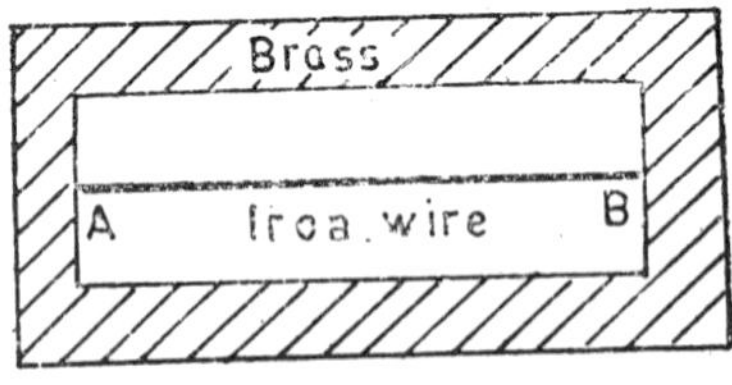

Fig. 2.29

Example 7:

It is required to fit a steel plug (diameter 0.1 m at 20°C) into a hole which has a diameter of 0.0999 m at 20°C Calculate the temperature to which the plug must be reduced to make this possible if α_{steel} = 0. 000011°C ^{1}.

Solution:

We have

$$l_{20} = 0.1 \text{ m} \quad l_t = 0.999 \text{ m}$$

$$\alpha = 0.000011°C^{-1} \text{ and } t = ?$$

From $\quad l_t = l_{20}[1 + \alpha (t - 20)]$

$\therefore \quad t = -71.0°C$

Example 8:

The pendulum of a clock is made of brass whose coefficient at linear expansion is 1.9×10^{-5} °C^{-1}. If the clock keeps correct time at 15°C, how many seconds per day it will lose at 20°C.

Solution:

For pendulums

$$T = 2\pi \sqrt{L/g}$$

$$\therefore \quad \frac{T_{20}^2}{T_{15}^2} = \frac{L_{20}}{L_{15}} = \frac{L_{15}(1 + \alpha . 5)}{L_{15}} = 1.000095$$

$$\therefore \quad \frac{T_{20}}{T_{15}} = 1.000047$$

$$\text{or} \quad \frac{T_{20} - T_{15}}{T_{15}} = 0.000047 \text{ seconds}$$

$$\text{or} \quad T_{20} - T_{15} = 0.0000477 \text{ sec.} \qquad [T_{15} = 1 \text{sec.}]$$

∴ Per day difference

$$= 24 \times 60 \times 60 \times 0.000047 = 4.1 \text{ seconds.}$$

Example 9:

(a) Using Tutton's compensator in Fizeaus' method tor determining the coefficient of linear expansion of a crystal it was found that when a crystal, 1.5 cm thick, Was heated through 40°C, 25 fringes crossed the field of view. If λ of the light used be 6000 Å, calculate α of the crystal.

(b) If a crystal has a coefficient of expansion. 13×10^{-7} in one direction and 231×10^{-7} in every direction at right angles to the first calculate its coefficient of cubical expansion.

Solution:

(a) We know

$$\alpha = \frac{p\lambda}{2l\,(t_2 - t_1)}$$

$$= 12.5 \times 10^{-6} \, {}^{\circ}C^{-1}$$

(b) The crystal here is clearly a uniaxial type with an axis of crystalline symmetry. The linear coefficients perpendicular to it are equal. So that

$$\alpha_x = 13 \times 10^{-7}{}^{\circ}C^{-1}$$

$$\alpha_y = \alpha_z = .231 \times 10^{-7}{}^{\circ}C^{-1}$$

$$\therefore \quad \gamma = \alpha_x + \alpha_y + \alpha_z$$

$$= 475 \times 10^{-7}{}^{\circ}C^{-1}.$$

Example 10:

If the coefficient of cubical expansion of glass and mercury are 2.5 $\times 10^{-5}$ and 1.8 $\times 10^{-4}$ respectively, what fraction of the whole volume of a glass vessel should be filled with mercury in order that the volume of the empty part should remain constant when glass and mercury are heated to the same temperature.

Solution:

Clearly, the volume of the empty part of the vessel will remain constant throughout, if the expansion in volume of mercury is just equal to the expansion in the volume of the vessel itself.

Let the initial volume of the vessel be V and of mercury be v, then

$$V_t = V(1 + \gamma_g t) \text{ or, } V_t - V = V\gamma_g t$$

and $$v_t = v(1 + \gamma_m t) \text{ or, } v_t - v = v\gamma_m t$$

but according to condition imposed

$$V\gamma_g t = v\gamma_m t$$

or $$\frac{v}{V} = \frac{\gamma_m}{\gamma_g} = \frac{5}{36}$$

Thus 5/36 of the total volume of the vessel must be filled with mercury.

Example 11:

(a) A thread of liquid, of expansion coefficient γ_l occupies l_0 length of a glass capillary tube of uniform area of cross-section, when the temperature is 0°C. What will be the length of the thread at t°C (i) if coefficient of linear expansion of glass be α, (ii) if

the expansion of the tube be negligible.

(b) What length of a glass tube, 98 cm long and closed at one end, should be filled with mercury so that the length unoccupied by mercury may remain the same at all temperatures? ($\gamma_g = 2.5 \times 10^{-5}$ and $\gamma_m = 1.8 \times 10^{-6}$).

Solution:

(a) (i) Let V_0 be the volume of the liquid thread at 0°C and a_0 be the area of cross-section of the tube then

$$l_0 = V_0/a_0$$

and $\quad V_t = V_0 (1 + \gamma_m t), \quad a_t = a_0 (1 + 2\alpha t)$

$$\therefore \quad l_t = \frac{V_t}{a_t} = \frac{V_0 (1 + \gamma_m t)}{a_0 (1 + 2\alpha t)}$$

$$= l_0 \frac{(1 + \gamma_m t)}{(1 + 2\alpha t)}$$

$$= l_0 (1 + \gamma_m t)(1 + 2\alpha t)^{-1}$$

$$\cong l_0 (1 + \gamma_m t - 2\alpha t)$$

i.e., true length of the liquid thread at t°C is

$$l_0[1 + (\gamma_m - 2\alpha)t]$$

(ii) In this case, since α is negligible

$$\therefore \quad l_t = l_0(1 + \gamma_m t)$$

(b) For the unoccupied length of the tube to remain the same at all temperatures increase in length of mercury column should be equal to increase in the length of the tube. Let l_0 be the occupied length of the tube at 0°C, then its increased length at t is

$$l_t = l_0[1 + (\gamma_m - 2\alpha)t]$$

or $\quad l_t - l_0 = l_0(\gamma_m - 2\alpha)t$

and increase in length of the tube $= \dfrac{l'\gamma_g t}{3}$

$$\therefore \quad \frac{l'\gamma_g}{3} = l_0 (\gamma_m - 2\alpha)\, t$$

where l' is the total length of the glass tube at 0°C and $\gamma_g/3$ is linear expansion of the tube.

Substituting proper values we get

$$l_0 = \frac{l'\gamma_g}{3(\gamma_m - 2\alpha)} = 5 \text{ cm}$$

i.e., a length of glass tube must be filled with mercury (Note–not necessarily at 0°C).

Example 12:

A little air has leaked into a barometer tube 1 m long. The mercury stands at 0.7 m when the tube is vertical and 0.78 m when it is inclined at 30° to the vertical. What is the atmospheric pressure.

Solution

In the first case when tube is vertical we have $P_v = H - 0.7$ and in the second case when the tube is inclined $P_i = H - 0.78 \cos 30 = H - 0.39\sqrt{3}$ where H is true barometric height. If 'a' be the area of cross-section of the tube then

$$V_v = (1 - 0.7)\, a$$

and $V_i = (1 - 0.39\sqrt{3})\, a$, then from Boyle's law

$$P_v V_v = P_i V_i$$

or, $(H - 0.7)(1 - 0.7)a = (H - 0.39\sqrt{3})(1 - 0.39\sqrt{3})a$

or $H = 0.7675$ m.

Example 13:

It is found that the volume of a certain gas increases in the ratio of 1:1.035 between 15°C and 25°C. Calculate the absolute zero on the celsius scale for this gas.

Solution:

Here pressure remains constant so

$$\frac{V_i}{V_F} = \frac{T_i}{T_F}$$

where i and F denote initial and final values of volume and temperatures for a given mass of a gas.

Let absolute zero is –x°C on the celcius scale,

then $T_i = (x + 15),\ T_F = (x + 25)$

$$\therefore \quad \frac{1}{1.035} = \frac{x + 15}{x + 25}$$

or, $x = -270.7°C$

Example 14:

Two equal bulbs are joined by a narrow tube and the system is filled with a gas at NTP and sealed. What will be the pressure of the gas if one of the bulbs is immersed in boiling water and the other in ice?

Solution

Let we have m kg of a gas whence

$$PV = m.r.T.$$

or $m = (PV_1/rT)$

at NTP both bulbs will have same amount of gas

$$\frac{m}{2} = \frac{0.76\ .\ V}{r\ .\ 273}$$

Now, in the second case gas from hot bulb will flow into the cold bulb till the pressure equals say p then mass of gas in hot bulb and cold bulb is

$$mH = \frac{p\ .\ V}{r\ .\ 373}$$

$$mC = \frac{p\ V}{r\ .\ 273}$$

But $mC + mH = m$

or $$\frac{pV}{r}\left[\frac{1}{373} + \frac{1}{273}\right] = \frac{2 \times 0.76 \times V}{r\ .\ 273}, \qquad [\because V_1 = 2V]$$

or $p = 0.8776$ m

Example 15:

A barometer made of very narrow tube is at NTP. Now temperature of the mercury is raised by 1°C ($\gamma_{mercury} = 0.00018°C^{-1}$) but *temperature of the atmosphere is not changed. What should be percent change in the height of mercury, if any?*

Solution:

The pressure of the atmosphere is constant, whence

$$H = L_0 d_0 g$$
$$= l_t \cdot d_t \cdot g$$

Hence, length of mercury column can increase at the cost of its density

$$l_t \cdot = l_0(1 + \gamma_m \cdot t)$$

or $$\frac{l_t}{l_0} - 1 + \gamma_m = 1.00018$$

$\therefore$ % change = 0.018.

EXERCISES

1. Two rods of different materials but of equal cross-sections (1 m length) are joined to make a rod of length 2 m. The metal of one rod has coefficient of linear thermal expansion 10^{-5} °C–1 and $Y = 3 \times 10^{+10}$ N m^{-2} and for another metal the values are 2×10^{-5} °C^{-1} and 10 N m^{-2}. How much pressure must be applied to the ends of the composite rod to prevent its expansion at temperature raised by 100°C.
2. An iron ring 1 m in diameter is to be put on a wooden cart wheel. The diameter of this wheel is 0.4 cm greater than that of the ring. By how much must the temperature of the ring be raised to fit it ($\gamma F_e = 1.2 \times 10^{-5}$ C^{-1})?
3. Two thin metal strips, one of brass and the other of iron, are bolted rigidly together, the strips being separated by washers 2 mm thick placed on each bolt If strips are of equal length at 20°C, calculate the radius of are formed by the compound strip when it is heated to 100°C?

   Given $\gamma_{brass} = 1.9 \times 10^{-5}$ °C^{-1}, $\gamma_{iron} = 1.2 \times 10^{-5}$ °C
4. Define the coefficient of linear expansion of a solid and describe so accurate method of determining it. The density of brass at 0°C is 8424 kg m^{-3} and its coefficient of linear expansion is 0.000018. Calculate density of brass at 40°C.

5. Describe how the coefficient of linear expansion of a solid can be determined with the help of an interferometric method.

6. Show how any three coefficients of linear thermal expansion for mutually perpendicular directions in a crystal are connected with the coefficient of volume expansion. How can a coefficient of linear expansion in the case of a crystal be satisfactorily determined by experiment?

7. An aluminium disc 8 cm in diameter at 15°C just fits a circular hole in a steel plate at 100°C. What is the area of the gap between them at 15°C if γ_{At}= 0.00007 °C^{-1} and γ_{steel} = 0.000012 °C^{-1}.

8. Increasing the temperature is one way of lengthening the rod and the other is by palling it. Find an expression for the force required to stop expansion, of the rod at t°C.

   Hint: Expansion of dl by heat is l a t and by force is dl = Fl/ay]

9. Expansion coefficients for liquids may be found (i) by change in the level of liquid, (ii) by finding mass of liquid at different temperatures, (iii) by hydrostatic upthrust on a solid immersed in a liquid, (iv) by balancing columns of hot and cold liquids. Are there methods on each of the said property? Give merits and demerits of each of the methods.

10. In order to annul the expansion of the vessel a quantity of mercury by volume V_0/μ is added to the vessel where Vo is the volume of the vessel. So that at t° volume of the vessel is $V_0(1 + ct)$ and of mercury V_0/μ $(1 + \mu ct)$ balance each other where 'c' is coefficient of vessel and μc is that of mercury. Show that since the capillary is also expanded the mercury level up to n division is also affected, whence real γ is

$$\gamma = \frac{(n - n_0)v_0}{V_0\left(1 - \frac{1}{\mu}\right)t}(1 + ct)$$

    where v_0 is the volume of one division of capillary at 0°C.

11. Define and distinguish between (i) the mean and zero expansion coefficients and (ii) apparent and absolute expansion coefficients of a liquid.

12. Describe Regnault's method of finding the absolute coefficient of expansion of mercury, explaining the precautions involved in it. Why is this method better than others?

13. Explain why a knowledge of the coefficient of expansion of mercury is considered to be so important.

    A glass bulb provided with a narrow stem graduated in hundred the of a c.c. is filled up to the zero mark on the stem by 10 cc of liquid at 0°C. If the coefficient of real expansion of liquid be seven times that of the glass, and the coefficient of linear expansion of glass is $8 \times 10^{-6}\ °C^{-1}$, calculate the reading at 40°C.

14. What conditions you will fix for a gas so that the two volume expansion coefficients are same for it.

15. Two weight thermometers are made of the same kind of glass. One is empty. Into the other a small piece of iron is introduced Describe how they may be used to determine the coefficient of expansion of iron.

16. A certain mass of gas occupies a volume of 5 litres at 0°C under a pressure of 0.2 m of mercury. Plot a graph showing a relation between pressure and volume at 0°C and at 273°C. Find from the graph the corresponding volumes at atmospheric pressure.

17. Describe Callendar's compensated air thermometer and show that it can be used to determine expansion coefficient of air.

18. At the sea level the barometer stands at 750 mm when the temperature is 7°C, while on the top of a mountain at –13°C it stands at 400 mm. Compare the weights of a cubic metre of air at the two places.

19. Benzene has a density of 0.9 gm/cc at 0°C and $\gamma = 0.0012\ °C^{-1}$. Balls of wood (density 0.88 gm/cc) float on it at 0°C If the wood has a coefficient of expansion of $4 \times 10^{-8}\ °C^{-1}$ along the grain and $4 \times 10^{-5}\ C^{-1}$ across the grain. Calculate at what temperature the ball will just sink.

20. How can you show that the density of water does not fall steadily as the temperature is raised from 0° to 10°C? When as it does (approximately) so beyond 10°C. Can you provide a feasible explanation?

21. In an experiment to determine coefficient of linear expansion of a crystal the following data was observed:

Thickness of crystal 0.5 cm, wavelength of light used 5154 Å. When the temperature of the crystal was increased from 20° to 65°C, the number of fringes that crossed the field of view was 12. Calculate the coefficient of linear expansion of the crystal.

22. A sphere of diameter 7 cm and mass 266.5 g floats in a bath of liquid. As the temperature is raised the sphere begins to sink at a temperature of 35°C If the density of the liquid is 1.527 g cm^{-2} at 0°C, find the coefficient of cubical expansion of the liquid. Neglect the expansion of the sphere.

23. A steel scale is to be prepared such that the millimetre intervals are to be accurate within 5×10^{-4} mm at a certain temperature. Determine the maximum permissible temperature variation during the ruling of the millimetre marks. For steel a is 13.22×10^{-6} $°C^{-1}$.

3

Calorimetry

3.1 DEFINITIONS

When a body is heated, its temperature rises. If 100g of copper and 100g of water are heated by similar burners for the same time, the rise in temperature is not the same in the two cases. The rise in temperature depends on the quantity of heat given to the body and the nature of its material. Let H be the quantity of heat given to a body of mass m and let the rise in temperature be θ.

Then, $$B = mC\theta$$

where C is a constant that depends upon the nature of the substance. C is called the specific heat of the substance.

Calorie : It is the quantity of heat required to raise the temperature of one gram of water from 14.5°C to 15.5°C. This is the standard unit recommended by International Committee of Pure Physics.

For ordinary purposes, the specific heat of water is taken as 1 but specific beat of water is not 1 at all temperatures. For practical purposes *Calorie* may be defined as the amount of heat required to raise the temperature of 1 gram of water through 1°C.

Kilogram Calorie : It is the amount of heat required to ruse the temperature of 1 kg of water through 1°C.

I kg calorie = 1000 calories = 1 kilocalorie.

British Thermal Unit : It is the amount of heat required to raise the temperature of 1 pound of water through 1°F.

1 B.T.U. = 252 calories

Therm. : It is the amount of heat required to raise the temperature of 10^5 pounds of water through IT.

1 Therm = 2-52 × 10^7 calories

Pound Calorie : It is the amount of heat required to raise the temperature of I pound of water through 1°C.

1 pound calorie = 453.6 calories

It is called centigrade heat unit.

Units of Heat

Unit	Quantity of water	Rise in tamp.	Relation
Calorie	1 g	1°C	= 1 Calorie
Kg Calorie	I kg	1°C	= 1000 Calories
BTU	1 pound	1°F	= 282 "
Therm	10^5 pounds	I°F	= 2.52 x 10^7 "
Pound Calorie	1 pound	1°C	= 453.6 "

Specific Heat : It is defined as the quantity of heat required to raise the temperature of unit mass of a substance through one degree.

Suppose,

Mass of the substance = m

Sp. heat of the substance = C

Rise in temperature = θ

$$H = mC\theta$$

or
$$C = \frac{H}{m\theta}$$

Unit of Specific Heat : In C.G.S. system, the unit of H is in calories, m is in grams and θ is in °C. Therefore the unit of specific heat will be

$$C = \frac{\text{Calorie}}{\text{g°C}}$$

Calorie/g°C

The specific heat of a substance is not constant and it is different at different temperatures. Ordinarily, the specific heat determined is the mean specific heat. Suppose, m is the mass of the substance, C the mean specific heat and if H units of heat is required to raise its temperature from θ_1 to θ_2, then the mean specific heat

$$C = \frac{H}{m(\theta_2 - \theta_1)}$$

For qualitative work, if dQ heat is given to raise the temperature of mg of a substance through $d\theta$,

$$dH = mCd\theta$$

$$C = \frac{1}{m} \cdot \frac{d\theta}{dt}$$

Thermal Capacity : It is the quantity of heat required to raise the temperature of the whole of the substance through 1°C. Let the mass of the substance be m and its specific heat C.

$$\text{Thermal capacity} = m \times C \times 1$$
$$= mC \text{ calories/°C}$$

Water Equivalent : It is the amount of water that will absorb the same quantity of heat as the substance for the same rise in temperature. Let the mass of the substance be m, specific heat C and rise in temperature θ.

$$H = mc\theta$$

If the water equivalent = w

$$H = w \times 1 \times \theta$$

$$\therefore \quad w \times 1 \times \theta = mC\theta$$

$$\text{or} \quad w = mC \text{ grams}$$

Water equivalent is numerically equal to the thermal capacity but the unit of water equivalent is grams and that of thermal capacity is calories.

3.2 REGNAULT'S METHOD OF MIXTURES

Specific Heat of Solids : Regnault was the first to devise the apparatus commonly used in the laboratories to find the specific heat of a solid or liquid employing the method of mixtures. In this the given solid is heated to a constant high temperature and then it is quickly transferred into a calorimeter containing water at room temperature. The final temperature of the mixture is noted. From the principle that heat lost by the solid is equal to the heat gained by the calorimeter and contents, the specific heat of the given solid is calculated.

3.3 COPPER BLOCK CALORIMETER

Specific heat of a solid at high temperature cannot be determined by ordinary apparatus used in the laboratory. Steam or vapours produced would cause considerable loss of heat when a hot solid is mixed with water or a liquid. The result obtained will not be accurate. To overcome this difficulty, *Nernst* and *Lindemann* designed a calorimeter called the *copper block calorimeter.*

It consists of a Dewar flask A having a copper block C fixed inside it. T, T are the terminals of the leads of the thermocouple to measure the temperature of the block C. The flask is covered from outside to prevent any loss of heat to the surroundings. Through the tube B, a hot solid can be dropped Fig. 3.1

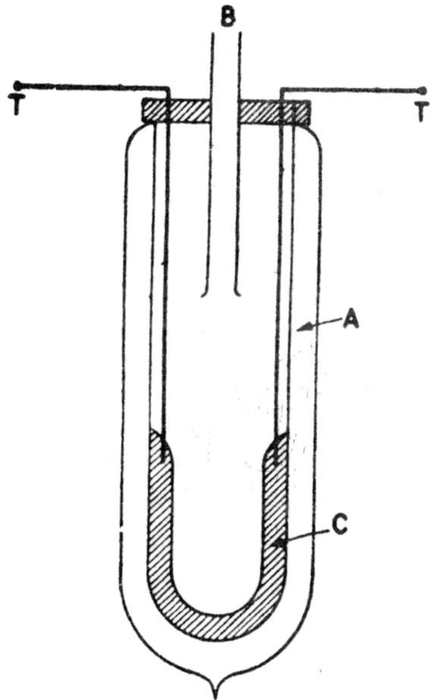

Fig. 3.1

The substance whose specific heat is to be determined is heated to a known high temperature and then gently dropped through the tube B into the block C. The copper block has high thermal conductivity and a uniform final temperature is reached in a short time. The temperature is measured with the help of a calibrated thermocouple, the copper block C acting as the hot end. To avoid air currents the tube B is closed with an automatic electrical arrangement after the substance has been dropped into the block C. The whole apparatus can be maintained at a desired temperature. From the heat gained by copper block and the heat lost by the substance, the specific heat of the substance can be calculated.

Magnus determined the specific heat of platinum and other metals up to 900°C. Jaeger and his co-workers determined the specific heat of P_t, tungsten Rh, Ir and other metals up to 1600°C. In order to find the specific heat at high temperatures, say between 700°C and 800°C, the experiment is performed twice, first at 700°C and then at 800°C. The mean value of the specific heat is calculated.

3.4 NERNST VACUUM CALORIMETER

This calorimeter is used to find the specific heat of solids at low temperature. Nernst and Lindemann performed a number of accurate experiments to determine the specific heat of solids at low temperatures.

The Nernst calorimeter consists of an evacuated flask D in which the metal whose specific heat is to be determined, is suspended as shown in Fig. 3.2. P and Q are two non-conducting loops. The cylinder B and the cylindrical plug A are made of the same metal. The heating coil C is made of platinum and is wound around A. The leads LL are connected to an electrical heating arrangement. The cylinder A fits completely inside the cylinder B. The heating coil C also serves as a platinum resistance thermometer to measure the temperature of the metal.

Suppose the resistance of the platinum wire is R_1, at temperature θ_1. R_1 is found by measuring the current in the wire and the potential difference across the two ends of the wire. Let the potential difference be E and the current I_1.

$$\therefore \qquad R_1 = \frac{E}{I_1} \text{ (At temperature } \theta_1\text{).}$$

Similarly, resistance R_2, at temperature θ_2 is given by

$$R_2 = \frac{E}{I_2} \text{ (At temperature } \theta_2\text{)}$$

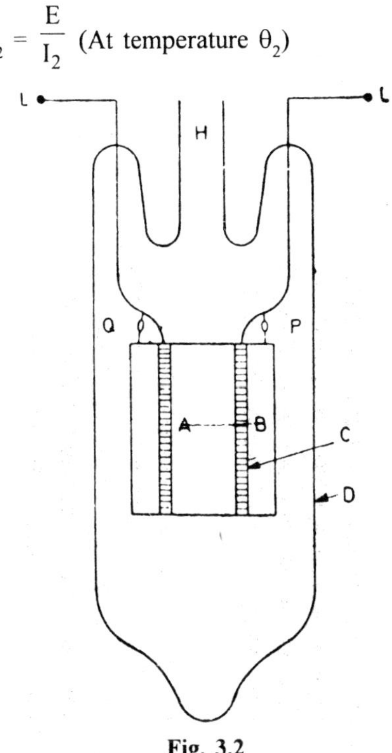

Fig. 3.2

The platinum wire is initially calibrated and from these values corresponding to R_1 and R_2 the values θ_1 and θ_2 are found.

The flask D is kept in a bath containing ice, liquid air etc., depending upon the low temperature at which the specific heat of the given metal is to be determined. After the flask D has attained the temperature of the bath, it is evacuated. This initial temperature of the metal is found. Let it be θ_1. Electric current is passed through the coil C for a known time so that the rise in temperature is about 1°C. Suppose the final temperature of the metal is θ_2.

Mass of the metal	= M grams
Specific heat	= C
Rise in temperature	= $(\theta_2 - \theta_1)$
Voltmeter reading	= E volts
Ammeter reading	= I amperes.

Time for which current is passed = t seconds

$$\therefore \quad MC(\theta_2 - \theta_1) = \frac{EIt}{4.2}$$

$$C = \frac{EIt}{4.2\ M\ (\theta_2 - \theta_1)}$$

In this experiment, the heat losses are practically eliminated as the experiment is performed in vacuum. However, a correction is required for the little loss of heat due to radiation.

To determine the specific heat of nonmetals, the substance is kept inside a silver vessel. The heating coil is wound round the silver vessel. The experiment is performed in vacuum by suspending the silver vessel inside the flask D.

Specific Heats and Atomic Heats of Solids at 20°C

Element	Specific Heat	Atomic Heat
Aluminium	0.212	5.72
Boron	0.307	3.32
Carbon	0.160	1.92
Copper	0.091	5.79
Gold	0.031	6.11
Iron	0.110	6.12
Lead	0.030	6.21
Silicon	0.182	5.11
Silver	0.058	6.04
Tin	0.054	6.31

3.5 NEWTON'S LAW OF COOLING

Newton's Law of Cooling states that the rate of loss of heat of a body is directly proportional to the difference of temperature of the body and the surroundings. The law holds good only for small difference of temperature. Also, the loss of heat by radiation depends upon the nature of the surface of the body and the area of the exposed surface.

$$\frac{dH}{dt} \propto (\theta - \theta_0) \text{ or } -\frac{dH}{dt} = k\ (\theta - \theta_0)$$

Consider a body of mass m, specific heat C and at temperature θ. Let θ_0 be the temperature of the surroundings. Suppose, the temperature falls by a small amount $d\theta$ in time dt. Then the amount of heat lost

$$dH = mCd\theta$$

$\therefore$ Rate of loss of heat

$$\frac{dH}{dt} = mC\frac{d\theta}{dt} \qquad ...(i)$$

From Newton's law of cooling

$$-\frac{dH}{dt} = k(\theta - \theta_0) \qquad ...(ii)$$

where k is a constant depending upon the area and the surface of the body.

From (i) and (ii)

$$-mC\frac{d\theta}{dt} = k(\theta - \theta_0)$$

or
$$\frac{d\theta}{\theta - \theta_0} = -\frac{k}{mC}.dt = -k.dt \qquad ...(iii)$$

Integrating, $\log(\theta - \theta_0) = -Kt + c$ ...(iv)

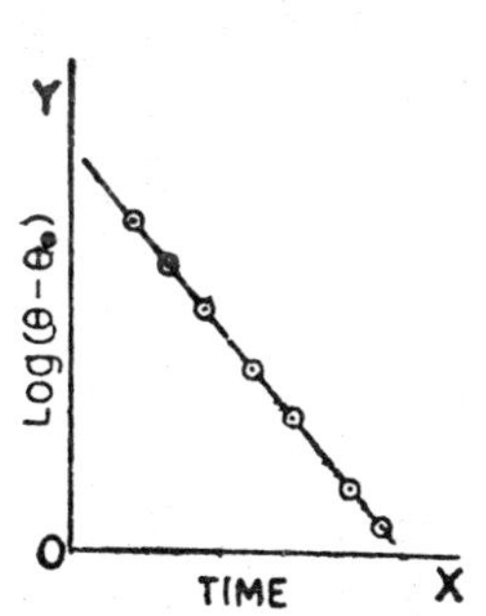

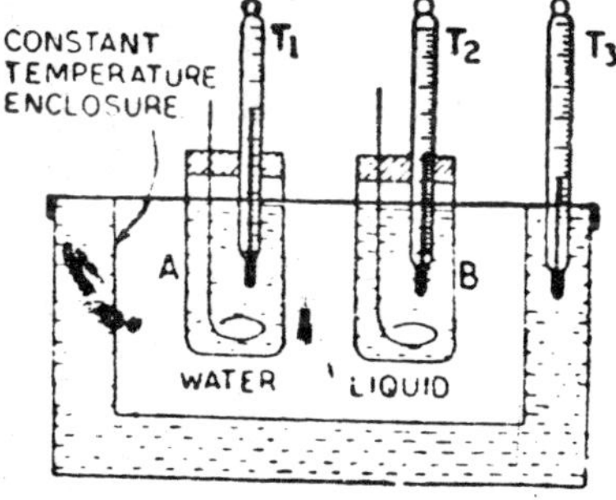

(i) (ii)

Fig 3.3

If a graph is plotted between t along the x-axis and log $(\theta - \theta_0)$ along the y-axis, it is a straight line. Hot water is taken in a calorimeter and is placed in a double walled vessel. Temperature of water after regular intervals of time is noted. A .graph between log $(\theta - \theta_0)$ and time t is plotted Fig. 3.3 (i). It is a straight line. This verifies Newton's law of cooling.

Specific Heat of a Liquid : A and B are two identical calorimeters containing equal volumes of hot water and the hot liquid respectively. The two calorimeters are made of the same material and their outer surfaces are equally polished. The calorimeters are kept inside a constant temperature enclosure. The thermometers T_1 and T_2 measure the temperature of water and liquid Fig. 3.3 (ii). The temperature of the two calorimeters are noted after regular intervals of time (say one minute). Graphs are plotted between temperature and time, for water and the liquid Fig. 3.4.

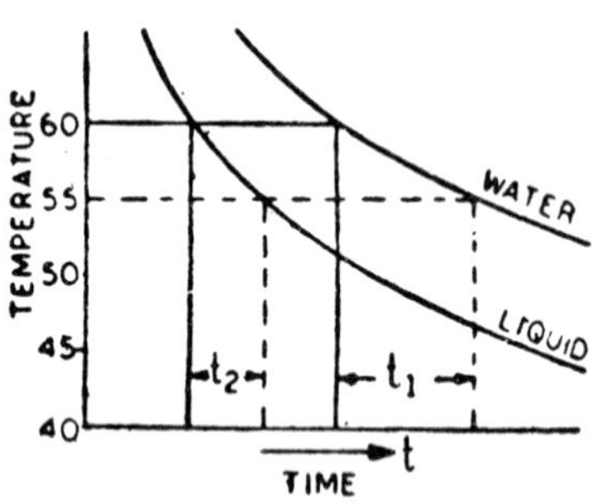

Fig. 3.4

From equation (iii)

$$\int \frac{d\theta}{\theta - \theta_0} = \frac{k}{mC} \int dt$$

In the case of water, suppose

Mass of water = m

Water equivalent of the calorimeter A = w

Time taken by water to cool from 60°C to 55°C = t_1

$$\therefore \quad \int_{60}^{55} \frac{d\theta}{\theta - \theta_0} = - \frac{-k}{(m + w)} \cdot t_1 \qquad ...(v)$$

Suppose mass of the liquid = M

Water equivalent of the calorimeter B = w

Specific heat of the liquid = C

Time taken by the liquid to cool from 60°C to 55°C = t_2

$$\therefore \quad \int_{60}^{55} \frac{d\theta}{\theta - \theta_0} = -\frac{k}{(MC + w)} \cdot t_2 \qquad \text{...(vi)}$$

From equations (v) and (vi)

$$\frac{MC + w}{t_2} = \frac{(m + w)}{t_1}$$

or

$$C = \frac{(m + w)t_2}{Mt_1} - \frac{w}{M} \qquad \text{...(vii)}$$

3.6 SPECIFIC HEAT OF A LIQUID—JOULES ELECTRICAL METHOD

The apparatus consists of a calorimeter in which a heater coil (wire) of resistance R is enclosed. The two ends of the wire are connected to the terminals on the lid. The calorimeter is enclosed in a wooden box. The wire R is connected in series with a battery, a key, a rheostat and an ammeter. A voltmeter is connected parallel to the wire (Fig. 3.5).

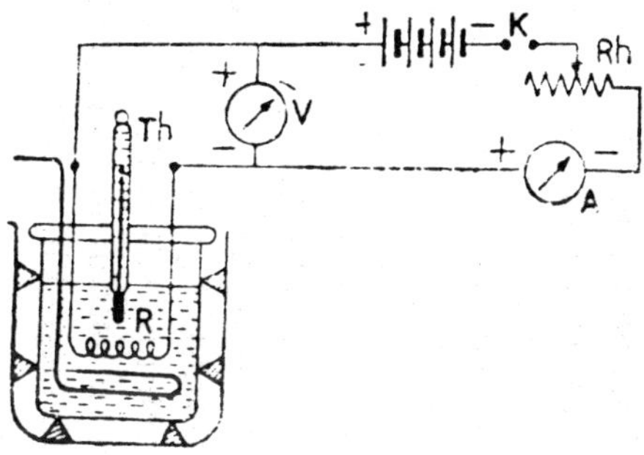

Fig. 3.5

The liquid whose specific heat is to be determined is taken in the calorimeter. Current is passed through the wire for a known interval of time. The rise in temperature of the calorimeter and the liquid is noted with the help of a thermometer. The current is passed for such a time that the rise in temperature is about 10°C.

Suppose, the mass of the liquid is M, specific heat of the liquid is C and water equivalent of the calorimeter is w.

Initial temperature of the liquid $= \theta_1$°C

Final temperature of the liquid $= \theta_2°C$

P.D. across the wire $= E$ volts

Current flowing $= J$ amperes

Time $=$ (seconds

Heat produced $= \dfrac{E.I.t}{4.2}$ calories

Heat gained by the liquid and the calorimeter

$$= (MC + w)\ (\theta_2 - \theta_1)$$

$$\therefore\ (MC + w) + (\theta_2 - \theta_1) = \frac{E.I.t}{4.2}$$

$$C = \frac{E.I.t}{4.2\ (\theta_2 - \theta_1)M} = \frac{w}{M}$$

3.7 SPECIFIC HEAT OF A LIQUID—CALLENDAR AND BARNES' CONTINUOUS FLAW METHOD

The apparatus consists of a glass tube in which a resistance wire is enclosed. The ends of the wire are connected to the terminals outside the tube. Thermometers T_1 and T_2 measure the temperature of the incoming and outgoing liquid. The liquid is passed through the tube at a uniform rate. The electrical connections are made as shown in Fig. 3.6. A vacuum jacket surrounding the glass tube is provided to avoid loss of heat by conduction and convection. Current and the rate of flow of water are adjusted so that the thermometers bow a difference of about 10°C.

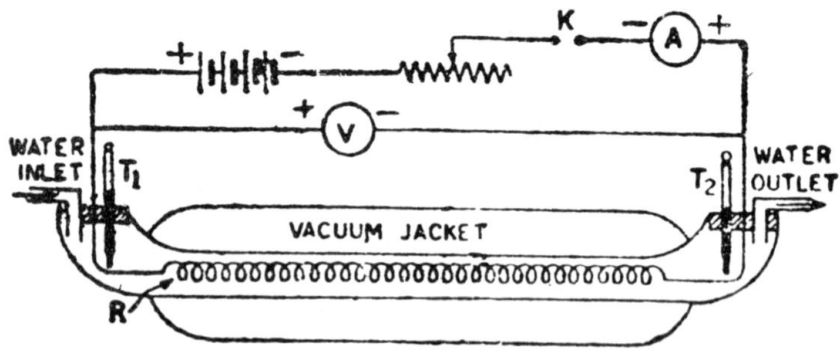

Fig. 3.6

Let, E, I and t be the potential difference, current and time for which the liquid is collected. If the mass of liquid collected is m and the difference of temperature between the two thermometers is $(\theta_2 - \theta_1)$, then

$$mC(\theta_2 - \theta_1) + R = \frac{E.I.t.}{4.2}$$

where R is the loss of heat by radiation. The current through the wire is changed. The rate of flow is adjusted so that the difference of temperature $(\theta_2 - \theta_1)$) remains the same. Suppose the amount of liquid collected in time t is m'

$$\therefore \qquad m'C(\theta_2 - \theta_1) + R = \frac{E.I.t.}{4.2} \qquad ...(ii)$$

Subtracting (ii) from (i)

$$(m - m')C\,(\theta_2 - \theta_1) = \frac{(EI - E'I')t}{4.2}$$

$$C = \frac{(EI - E'I')t}{4.2\,(m - m')(\theta_2 - \theta_1)}$$

This is an accurate method because the observations are taken under steady state. The thermal capacity of the apparatus does not occur in the calculations. Moreover, the errors due to loss of heat by conduction, convection and radiation have been practically eliminated.

3.8 EXPERIMENTAL DETERMINATION OF HEAT CAPACITIES

In modern methods, measurement of heat capacity of substances involves the supply of heat to the system from an electrical source. From the energy received from the source in a given interval of time, a part of it may be used in raising the temperature of the system and the remainder is lost to the surroundings at a lower temperature (Fig. 3.7).

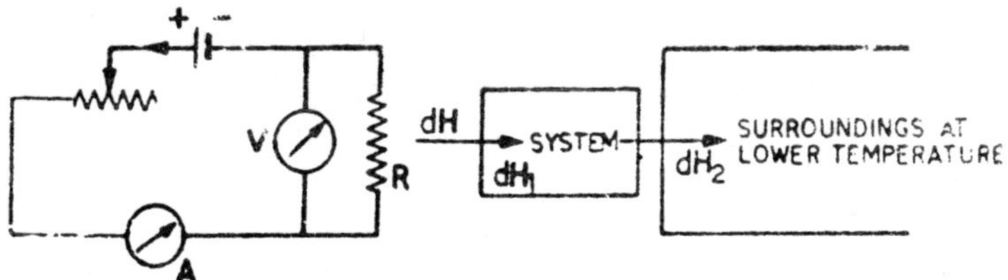

Fig. 3.7

Let dH be the quantity of heat drawn from the electrical source in time dt. Here dH_1 is the amount of heat retained by the system to raise its temperature by $d\theta$ and dH_2 is the amount of heat lost to the surroundings.

$$dB = dH + dH_2$$

$$dH_1 = 1 \times C_p \times dT$$

Here C_p is the specific heat of a substance at constant pressure. If the potential difference across the heater wire is E volts and the current flowing is I amperes, then

$$dH = \frac{EI.dt}{J}$$

$$\therefore \quad dH_1 = dH - dH_2$$

$$= \quad C_p \times dT = \frac{EI.dt}{J} - dH_2 \qquad ...(i)$$

If the system does not lose heat to the surroundings, then

$$dH_2 = 0$$

$$\therefore \quad C_p \times dT = \frac{EI.dt}{J} \qquad ...(ii)$$

$$\frac{EI}{J} = C_p \left[\frac{dT}{dt}\right] \qquad ...(iii)$$

To find dT/dt, *i.e.*, the rate of rise of temperature of the system with time, a curve is drawn between temperature (T) and time (t). From the graph dT/dt is found for the temperature at which 0, is to be determined.

3.9 ADIABATIC VACUUM CALORIMETER

This calorimeter was designed by *Nernst* and later on modified by *Simon* and *Langc*. Inside a thin copper vessel C, the substance is taken. The vessel C also contains heating coils and a sensitive platinum resistance thermometer to note the temperature. A thermostat B is made of copper and surrounds the vessel C. It is also heated electrically. The two junctions of the sensitive thermocouple are in contact with the inner surface of B and the outer surface of C.

The outer vessel A is surrounded by liquid hydrogen contained in a vacuum flask. The current through the heater coil of B is adjusted so that the thermocouple shows no deflection: This ensures a uniform temperature enclosure for C.

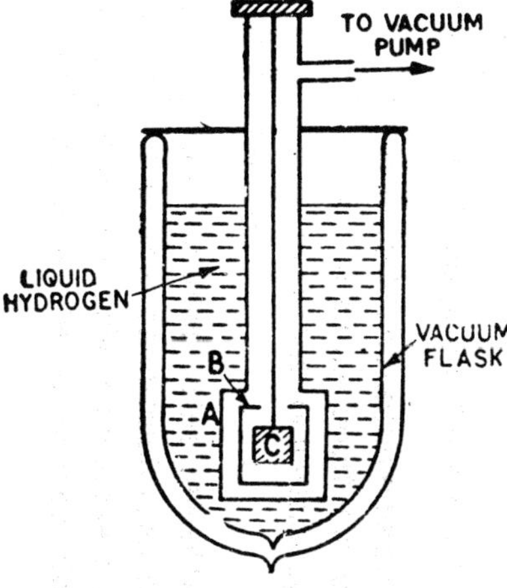

Fig. 3.8

The space inside A is evacuated with the help of a vacuum pump. If the rise in temperature of 1 gram of the substance is dT in time dt.

Then $$\frac{EI}{J} = C_p \left[\frac{dT}{dt}\right] \qquad \text{...(iv)}$$

The method is useful in determining the heat capacity of substances at low temperatures.

To find the value of C_p, it is difficult to find its value experimentally at low temperature. Its value is calculated from the equation

$$C_v = C_p - \frac{r}{J} \qquad \text{...(v)}$$

3.10 TWO SPECIFIC HEATS OF A GAS

Consider a gas of mass mat a pressure P and volume F. If the gas is compressed, there is rise in temperature. In this case, no heat has been supplied to the gas to raise its temperature.

$\therefore$ Specific heat, $C = \frac{H}{m\theta}$

But $\qquad H = 0$

$\therefore \qquad C = 0$

On the other hand, if heat is supplied to the gas and the gas is allowed to expand such that there is no rise of temperature, then

$$C = \frac{H}{m\theta}$$

Here $\qquad \theta = 0$

$$\therefore \qquad C = \frac{H}{m \times 0} = \infty$$

Thus, the specific heat of a gas varies from zero to infinity.

In order to fix the value of the specific heat of a gas, the pressure or volume has to be kept constant. Consequently, a gas has two specific heats.

(1) Specific heat at constant volume C_v,

(2) Specific heat at constant pressure C_p.

C_v: It is defined as the quantity of heat required to raise the temperature of one gram of a gas through 1°C at constant volume.

C_p: It is defined as the quantity of heat required to raise the temperature of one gram of a gas through 1°C at constant pressure.

C_p is greater than C_v: When a gas is heated at constant volume, the heat supplied to the gas is wholly used up to raise its temperature. On the other hand when a gas is heated at constant pressure, a part of the heat is used to raise its temperature and a part is used to do external work to keep the pressure constant.

$$C_p > C_v$$

Relation : Consider one gram of a gas at a pressure P, volume V and temperature T. Heat is supplied to the gas to raise its temperature through dr. As the pressure has to remain constant,

Work done, $W = P \times A \times x = P \times dV$

where dV is the change In volume.

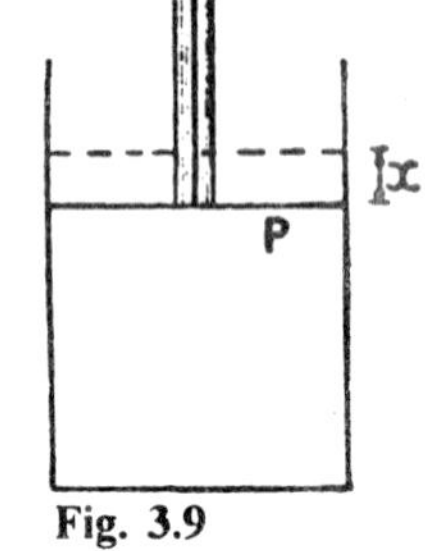

Fig. 3.9

From the gas equation

$$PV = tT,$$

Differentiating,

$$PdV + VdP = r\ dT$$

But $dP = 0$

$\therefore$ $PdV = rdT$

$\therefore$ Work done in heat units

$$= \frac{r.dT}{J} \text{ calories}$$

Heat supplied $= 1\ 1\ C_p \times dT$

$$= 1 \times C_v\ dT + \frac{r.dT}{J}$$

or $C_p - C_v = r/J$

where r is the gas constant for one gram of a gas. If C_p and C_v represent gram molecular specific heats, then

$$C_P - C_V = \frac{R}{J}$$

where R is the universal gas constant.

3.11 SPECIFIC HEAT OF A GAS AT CONSTANT VOLUME—JOLLY'S DIFFERENTIAL STEAM CALORIMETER

C is a chamber in which steam can be admitted. P_1 and P_2 are scale pans suspended from the scale pans S_1 and S_2 of a balance.

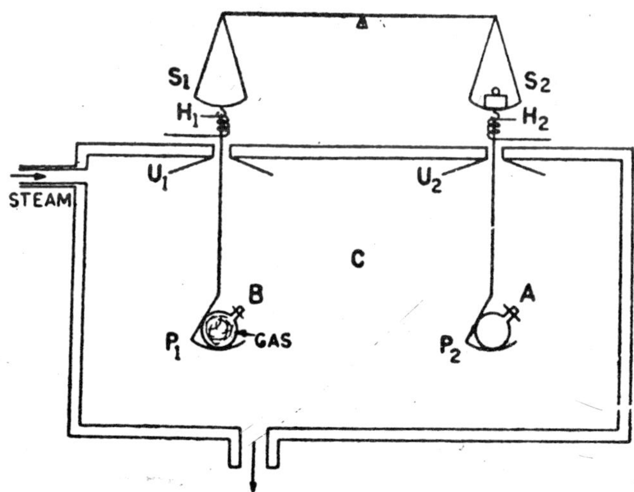

Fig. 3.10

A and B are two identical hollow metal spheres. The two spheres are initially evacuated and by placing them in the pans P_1 and P_2 the balance is counterpoised. In one of the spheres, the gas whose C_v is to be determined is admitted at a pressure of about 10 atmospheres and the balance is counterpoised again. Suppose the gas is filled in B. The balance will tilt to the left and the extra mass that has to be kept in the pan S_2 for counterpoising corresponds to that mass of gas enclosed in B (M grams). Now steam is admitted into the chamber and passed continuously till a constant temperature is reached (temperature of steam). Steam condenses and the mass of steam that has condensed on the pan P_1 is more than that condensed on the pan P_2 because in the case of pan P_1 the enclosed gas also has to be heated from the room temperature to the temperature of steam.

The balance tilts and an extra mass has to be kept in the right hand pan to balance it again. This extra mass corresponds to the extra mass of steam condensed on the pan P_1.

The umbrella-shaped vanes U_1 and U_2 will not allow the steam condensed on the rest of the chamber to fall on the scale pans P_1 and P_2. The heating coils H_1 and H_2 will heat the suspension wires to a temperature higher than the temperature of steam and therefore, no steam condenses on these wires.

Suppose,

Mass of the gas = M grams

Sp. heat of gas at constant volume = C_v

Initial temperature of the gas = t_1°C

Temp. of steam = t_1°C

Extra mass of steam condensed on the scale pan P_1 = m grams

Latent heat of vaporisation of water = L cals/g

Heat lost by steam = mL

Heat gained by the gas = $MC_v(t_2 - t_1)$

$$MC_v(t_2 - t_1) = mL$$

$$C_v = \frac{mL}{M(t_2 - t_1)} \text{cals/g}^{o}\text{C}.$$

In the second part of the experiment the gas is enclosed in the sphere of pan P_2 and the sphere of pan Pi is kept empty and C_v is calculated again. The mean of these two values gives the specific heat of the gas at constant volume.

3.12 SPECIFIC HEAT OF A GAS AT CONSTANT PRESSURE—(REGNAULT'S METHOD)

The apparatus consists of a reservoir R containing the gas at high pressure and at a constant temperature. The pressure of the gas in the reservoir is shown by the pressure gauge. The apparatus is as shown in Fig. 3.11.

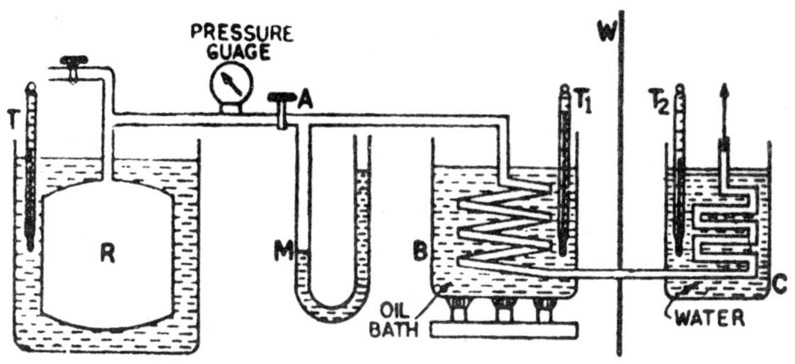

Fig. 3.11

The regulator A allows the gas to flow at a constant pressure through the spiral tubings immersed in the oil bath B and the calorimeter. The pressure of the gas flowing through the spiral tubing is shown by the manometer M. Regulator A helps in keeping the level of the liquid in the manometer limbs constant.

Suppose the initial pressure of the gas at any instant in the reservoir is P_1, its temperature is T_1 and volume is V. The temperature shown by the oil bath is T_1 and the calorimeter is at a temperature T_2. Gas is allowed to flow for about half an hour. The gas after passing through the oil bath gets heated to temperature T_1 and after passing through the calorimeter, gets cooled and gives heat to the calorimeter and its contents. Suppose, the final pressure of the gas in the reservoir is P_2 volume is V and the temperature of the calorimeter C and its contents is T_3.

Calculations

Suppose, mass of the gas flown = M

Mass of water in the calorimeter = m

Water equivalent of the calorimeter = w

Rise of temperature of calorimeter and its contents

$$= (T_3 - T_2)$$

Mean fall of temperature of the gas

$$= \left(T_1 - \frac{T_2 + T_3}{2}\right)$$

Heat gained = Heat lost

$$(m + w)(T_3 - T_2) = MC_p = \left(T_1 - \frac{T_2 + T_3}{2}\right)$$

$$C_p = \frac{(m + w)(T_3 - T_2)}{M\left(T_1 - \frac{T_2 + T_3}{2}\right)} \quad \text{...(i)}$$

To find the mass of the gas (M), suppose the density of the gas at NTP = ρ. In the experiment, Vcc of the gas at a pressure $(P_1 - P_2)$ and temperature TK has flown through the apparatus. Reducing the volume of the gas to NTP,

$$\frac{(P_1 - P_2)V}{T} = \frac{76\ V_0}{273}$$

or $$V_0 = \frac{(P_1 - P_2)V \times 273}{76 \times T}$$

$$\therefore \text{ Mass } M = V_0 \times \rho = \frac{(P_1 - P_2)V \times 273 \times \rho}{76 \times T}$$

Thus, knowing the value of M, C_p can be calculated from equation (i).

3.13 CONTINUOUS FLOW ELECTRICAL METHOD

The specific heat of a gas at constant pressure by electrical method can be determined by using Callendar and Barnes' continuous now apparatus. Here D is a special glass vessel. The heating coil is at the axis of the vessel. The incoming gas takes a long zig-zag path as indicated by the arrow heads (Fig. 3.12).

The reservoir B contains the gas at high pressure and at a constant temperature. The pressure of the gas is read by the pressure gauge. The regulator A allows the gas to flow at a constant pressure through the vessel D. The pressure of the gas flowing through D is indicated by the manometer M. The regulator helps in keeping the pressure of the gas constant throughout the experiment. The filament is heated by electric current. The incoming gas is heated due to the heat generated by the filament. The inlet and the outlet temperatures of the-gas are measured with the help of the platinum resistance thermometers T_1 and T_2.

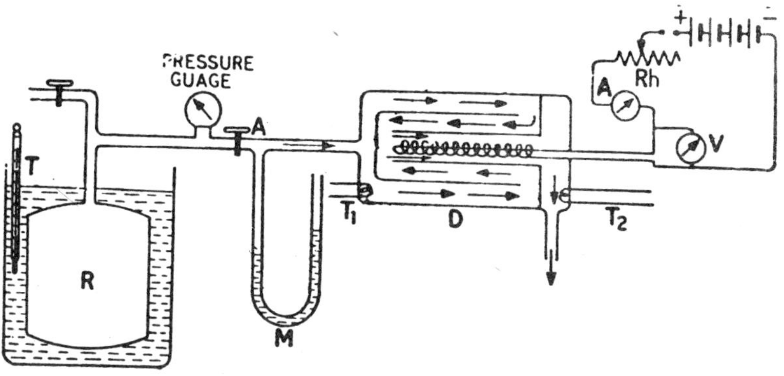

Fig. 3.12

The gas is allowed to flow through the apparatus for some time till the steady state is reached. When the steady state is reached, the thermometers T_1 and T_2 show constant readings. After the steady state, note the pressure (P_1) of the gas in the reservoir. Allow the gas to flow for half an hour and note the final pressure (P_2) of the gas in the reservoir. During this half an hour, the manometer M and the thermometers T_1 and T_2 should show constant readings.

Calculations

Suppose, the mass of the gas flown	$= M$
Temperature of incoming gas	$= T_1$
Temperature of outgoing gas	$= T_2$

Specific heat of the gas at constant pressure	$= C_p$
Voltmeter reading	= E volts
Ammeter reading	= I amperes
Time	= t seconds
Heat produced	$= \frac{EIt}{4.2}$ calories
Heat gained by the gas	$= MC_p(T_2 - T_1)$.

Heat gained = Heat produced

$$MC_p\ (T_2 - T_1) = \frac{EIt}{4.2}$$

$$C_p = \frac{EIt}{4.2\ M\ (T_2 - T_1)} \qquad \text{...(i)}$$

To find the mass of the gas (M), suppose the density of the gas at N.T.P. = ρ. In the experiment Vcc of the gas at a pressure $(P_1 - P_2)$ and temperature T has flown through the apparatus. Here V is the volume of the reservoir R. Reducing the volume of the gas to N.T.P.

$$\frac{(P_1 - P_2)\ V}{T} = \frac{76\ V_0}{273}$$

$$V_0 = \frac{(P_1 - P_2)\ V \times 273}{76T}$$

$$\therefore \text{ Mass } M = \rho V_0 = \frac{\rho(P_1 - P_2)\ V \times 273}{76T} \text{ grams}.$$

Thus, knowing the value of M, C_p can be calculated from equation (i).

The chief advantages of this method are:

1. The temperature of every portion of the apparatus remains constant when the steady state is reached.
2. The thermal capacity of the flow tube and its contents is eliminated in the calculations.
3. The loss of heat due to radiation is minimised due to zig-zag path taken by the gas. The gas flows from outer to the inner region of the tube and absorbs any heat radiated from the inner to the outer portion.

4. The temperatures can be measured accurately under steady state.
5. The heat produced by electrical arrangement can be calculated accurately.

3.14 SPECIFIC HEAT OF A GAS AT LOW TEMPERATURES

The continuous flow method can be used to determine the specific heat of a gas at low temperature. Before the gas is passed into the tube D, the gas is initially cooled fo a desired low temperature. The experiment is performed in a similar way as discussed in article 3.13.

This method is also used to find the specific heat of a gas at different high pressures. In this case, the gas at high pressure is passed through the tube D. Scheel and Heuse employed this method to find the specific heat of various gases up to –180°C. Holborn and Jakob used this method to find the specific heat of gases at high pressures.

Molecular Heats of Gases at 20°C

Gas	C_p	C_v
Hydrogen	6.87	4.88
Nitrogen	6.95	4.90
Oxygen	7.03	5.04
Chlorine	8.29	6.15
Air	6.95	4.96
Carbon dioxide	8.83	6.80
Helium	4.97	2.98
Argon	4.97	2.98

The specie heats of hydrogen (for 1 gram) are

$$C_p = 3.435 \text{ and } C_v = 2.44.$$

3.15 CALORIFIC VALUE OF FUELS

Calorific value of a fuel is defined as the quantity of heat released when a unit quantity of the fuel is completely burnt and the products of combustion are brought to the original temperature. It is expressed as calories per gram or B. T. U. per pound. The fuels commonly used are coal, petrol, spirit, diesel oil, wood etc. The calorific value of different fuels can be determined with the help of a bomb calorimeter.

Bomb Calorimeter

It consists of a vessel D made of steel or gun metal. A platinum bowl B is suspended inside D. The fuel E in the form of powder for a solid or a liquid is taken in B. A heating wire is immersed in the fuel and its ends are connected to the terminals T_1 and T_2 R is a regulating valve for the free supply of oxygen inside D. The vessel D is closed with an air tight lid. The whole apparatus is immersed in a calorimeter C containing water. The fuel is ignited by passing current through the heater wire. The heat produced due to the combustion of the fuel is taken by the surrounding water in the calorimeter.

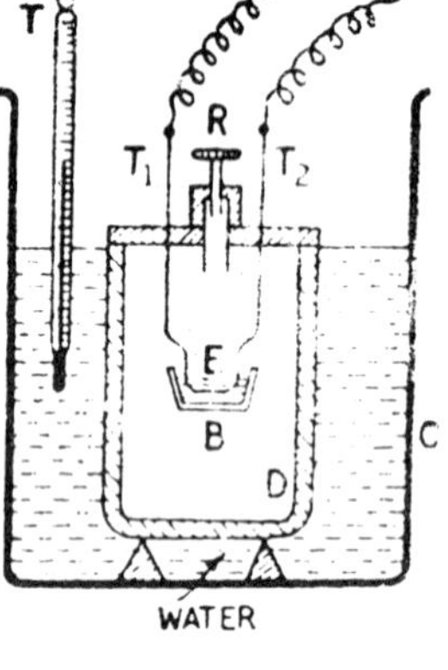

Fig. 3.13

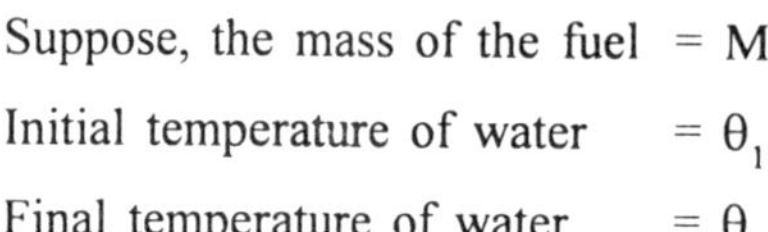

Suppose, the mass of the fuel = M

Initial temperature of water = θ_1

Final temperature of water = θ_2

Mass of water = m

Water equivalent of the calorimeter and the vessel D = w

Heat produced $= (m + w)(\theta_2 + \theta_1)$

$\therefore$ Calorific value $= \dfrac{(m+w)(\theta_2 - \theta_1)}{M}$

3.16 BELL CALORIMETER

It is used to find the calorific value of fuels. The apparatus consists of a bell jar B arranged inside a calorimeter containing water at room temperature. In the crucible C, the substance is taken in the powdered form. The leads L, L are connected to a heating filament dipped inside the fuel in the crucible 0 (Fig. 3.14). D is a perforated disc through which the burnt gases inside the bell can escape into water. A is a pipe through which oxygen is fed into the combustion chamber.

The fuel is ignited by connecting the leads L, L to a battery. The fuel burns and the burnt hot gases escape into water in the calorimeter, through the disc D. When the fuel is completely burnt, the final temperature of water in the calorimeter is noted.

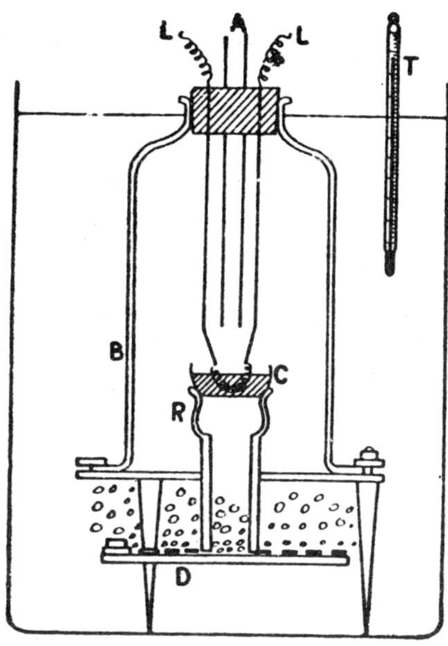

Fig. 3.14

Suppose, the mass of the fuel $= M$

Initial temperature of water $= \theta_1$

Final temperature of water $= \theta_2$

Mass of water $= m$

Water equivalent of the calorimeter $= w$

Heat produced $= (m + w)(\theta_2 + \theta_1)$

$\therefore$ Calorific value $= \dfrac{(m+w)(\theta_2 - \theta_1)}{M}$

Calorific Value of Fuels

Fuel	Cal/g
Wood	2500
Gas coke	6000
Methylated spirit	6400
Steam coal	7500
Anthracite coal	8800
Heavy Diesel oil	11350
Petrol	11400
Paraffin oil	11200

3.17 DULONG AND PETIT'S LAW

Dulong and Petit, in 1819, studied the specific heat of various elements in a solid state and enunciated a law, called Dulong and Petit's law. According to this law, the *product of the specific heat and the atomic weight i.e., atomic heat of all the elements in the solid state is a constant.* The value of this constant was fixed as 6.4 but it is taken as 6 at present. The exact value is 5.96 which also agrees with the value derived from the kinetic theory of matter.

The justification of Dulong and Petit's law was obtained from Boltzman's consideration of the law of equi-partition of energy. According to it, the energy associated with one gram atom of a substance for each degree of freedom at temperature T = 1/2 RT. Here R is the universal gas constant. If the atom is considered to be vibrating about the mean position, its mean kinetic energy will be equal to its mean potential energy.

For each form of energy there are three degrees of freedom. Therefore an atom has got six degrees of freedom. Thus, the total energy associated with one gram atom of a substance at a temperature T = 3RT.

$$\therefore \quad U = 3RT$$

$$A_H = \frac{dU}{dT} = 3R$$

A_H is the atomic heat of the substance

$$A_H = \frac{3 \times 8.31 \times 10^7}{4.8 \times 0^7} \text{ cals/g} - \text{atom} - \text{K}$$

$$A_H = 5.96 \text{ cals/g-atom-K}$$

Atomic Heat of Substances at 20°C

Substance	Atomic Weight	Specific Heat	Atomic Heat
Aluminium	27.0	0.212	5.72
Boron	10.8	0.307	3.32
Carbon	12.0	0.160	1.92
Copper	63.6	0.091	5.79
Gold	197.2	0.031	6.11
Iron	55.8	0.110	6.12
Lead	207.2	0.030	6.21
Silicon	28.1	0.182	5.11
Silver	107.9	0.056	6.04
Zinc	65.4	0.092	6.02

Dulong and Petit's law was modified by Woestyn for metallic compounds. According to him, the *molecular heat of* a compound is equal to the sum of the atomic heat of its constituents. Further, it was found by Newmann that the molecular heat of the compounds of similar nature is also constant. The molecular heat in the case of NaCl, AgCl and KCl was found to be nearly 13 and that of Sb_2O_3, Fe_2O_3, As_2O_3 was found to be equal to 26.

3.18 VARIATION OF SPECIFIC HEAT AND ATOMIC HEAT WITH TEMPERATURE

It is found that Dulong and Petit's law is not true in the case of carbon, boron and silicon. In the case of these elements the atomic heats at 20°C are 1.92, 3.32 and 5.11. These values differ from the constant value of 6. This variation in atomic heat could not be explained on the basis of kinetic theory of matter. However, it was found by Nernst that the specific heat of a substance decreases with decrease in temperature and at absolute zero the specific heat tends to be zero. Further, he was able to show that the specific heat increases with the rise in temperature and tends to a maximum value. Therefore, the atomic heat of a substance tends to a maximum value of six and decreases with decrease in temperature. In the case of carbon, boron and silicon also, the atomic heat is 6 at high temperatures (Fig. 3.15).

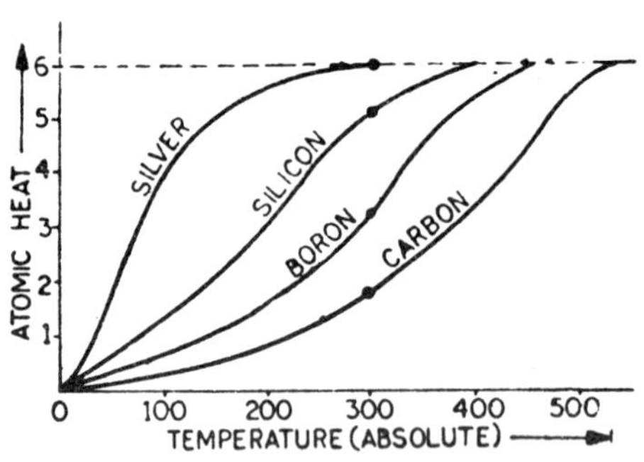

Fig. 3.15

In the case of silver, the atomic heat is 6 at room temperature but it is also less than 6 at lower temperatures.

Atomic Heats of Silver at Low Temperature

Temperature (K)	Atomic Heat
205	5.61
144	5.37
103	4.80
75	4.04
56	3.19
36	1.69
20	0.40
10	0.05
5	0.005
1.4	0.00025

3.19 QUANTUM THEORY

Dulong and Petit's law has been explained on the basis of quantum theory of heat radiation. According to quantum theory, heat is radiated in the form of discrete particles called photons. Each particle has energy equal to Ay where h is the Planck's constant and v is the frequency of heat radiations. Einstein also explained Dulong and Petit's law on the basis of quantum theory and said that the atomic heat is equal to 6 only at higher temperatures, and this is the maximum value. The atomic heat of the elements decreases with decrease in temperature. Most of the substances reach the maximum value of six at room temperature but carbon, born and silicon are below the maximum value of six at room temperature. Their atomic heats also tend to the maximum value of 6 at higher temperature.

Debye also modified Einstein's theory and concluded that in some substances like copper, aluminium, iron etc., the atomic heat at low temperature decreases more slowly than explained by Einstein.

According to Debye-Einstein Theory:

1. The atomic heat of all the elements tends to the maximum value of six.
2. The atomic heat of all the elements decreases with the fall of temperature.
3. The atomic heat tends to the value zero, near about absolute zero.
4. At low temperatures, the atomic heat varies as the cube of the absolute temperature.

   $$\text{Atomic heat } \alpha\ T^3$$

   It is known as Debye's T^3 law.
5. The graphs between atomic heats and absolute temperature of all the elements will coincide if the temperature scale is suitably modified according to the expression,

   $$A = f\left(\frac{\theta}{T}\right).$$

Here, A is the atomic heat of the element at absolute temperature T and 9 is a parameter called Debye temperature. Debye temperature is a constant for a particular element but is different for different elements.

The value $f\left(\frac{\theta}{T}\right)$ is the same for all the elements at absolute temperature T. Thus, according to this theory a single graph between. A and $f\left(\frac{\theta}{T}\right)$ for all the elements is obtained (Fig. 3.16).

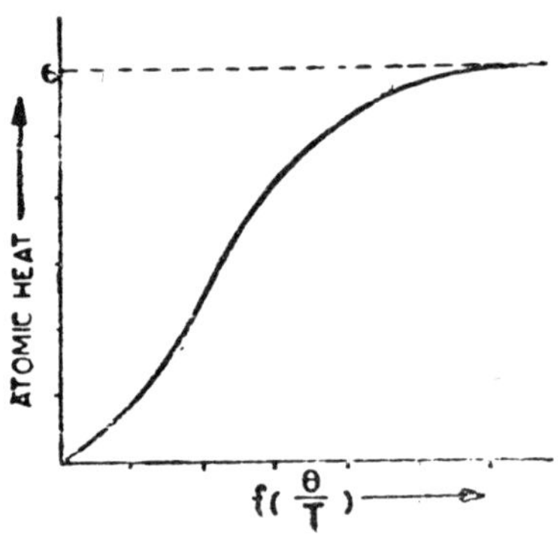

Fig. 3.16

SOLVED EXAMPLES

Example 1:

A liquid takes 5 minutes to cool from 80°C to 50°C. How much time will it take to cool from 60°C to 60°C? The temperature of the surrounding is 20°C.

Solution:

We have $\int \frac{d\theta}{\theta - \theta_0} = -K \int dt$

$$\theta_0 = 20°C$$

In the first case,

$$\int_{80}^{50} \frac{d\theta}{\theta - \theta_0} = -K \times 5$$

$$\log \left[\theta - \theta_0\right]_{80}^{50} = 5K$$

$$\log\left(\frac{50-20}{80-20}\right) = -5K$$

$$\log\frac{30}{60} = -5K$$

or $$\log\frac{60}{30} = 5K$$

In the second case, suppose the time taken is t minutes,

$$\int_{60}^{30}\frac{\theta}{\theta-\theta_0} = -K$$

$$\log\left(\frac{30-20}{60-20}\right) = -Kt \qquad \text{...(i)}$$

$$\log\frac{10}{40} = -Kt$$

or $$\log\frac{40}{10} = -5t \qquad \text{...(ii)}$$

Dividing (ii) by (i)

$$\frac{t}{5} = \frac{\log 4}{\log 2} = 2$$

t = 10 minutes.

Example 2:

Equal volumes of water (density 1 g/cm^3) and alcohol (density 0.8 g/cm^3) when put in similar calorimeters take 100 seconds and 74 seconds respectively to cool from 50°C to 40°C Calculate the specific heat of alcohol. Thermal capacity of each calorimeter is numerically equal to the volume of either liquid.

Solution:

Let the volume of either liquid be V

Mass of water = m = V × 1

Mass of alcohol = M × V × 0.8

Water equivalent of each calorimeter = w = V

Here t_1 = 100s, t_2 = 74s

$$\frac{MC + w}{t_2} = \frac{m + w}{t_1}$$

$$C = \frac{(m + w)t_2}{Mt_1} - \frac{w}{M}$$

$$= \frac{(V + V)74}{V \times 0.8 \times 100} - \frac{V}{V \times 0.8}$$

C = 0.6 calorie/g–K.

Example 3:

A body cools in 8 minutes from 60°C to 40°C. What will be its temperature after the next 5 minutes? Temperature of the surroundings = 10°C. Assume that the Newton's law of cooling holds good throughout the process.

Solution:

We have $\int_{\theta_1}^{\theta_2} \frac{d\theta}{\theta - \theta_0} = -K\int dt = -Kt$

(1) In the first case, $\theta_0 = 10°C$,

$\theta_1 = 60°C$, $\theta_2 = 40°C$, $t = 5$ minutes

$$\int_{60}^{40} \frac{d\theta}{\theta - \theta_0} = -5K$$

$$\left[\log(\theta - \theta_0)\right]_{60}^{40} = -5K$$

$$\log\left(\frac{40 - 60}{60 - 10}\right) = -5K \qquad \text{...(i)}$$

(2) In the second case

$\theta_0 = 10°C$

$\theta_1 = 40°C$, $\theta_2 = x$, $t = 5$ minutes

$$\int_{40}^{x} \frac{d\theta}{\theta - \theta_0} = -5K$$

$$\log\left(\frac{x-10}{40-10}\right) = -5K \qquad ...(ii)$$

From equations (i) and (ii)

$$\log\left(\frac{x-10}{40-10}\right) = \log\left(\frac{40-10}{60-10}\right)$$

$$\frac{x-10}{30} = \frac{30}{50}$$

$$\mathbf{x = 28°C.}$$

Example 4:

A liquid fakes 4 minutes to wolfram 70°C to 60°C. How much time will it take to cool from 50°C to 40°C? The temperature of the surroundings is 25°C. Newton's law of cooling is applicable throughout the process.

Solution:

We have $\int_{\theta_1}^{\theta_2} \frac{d\theta}{\theta - \theta_0} = -K\int dt = -Kt$

(1) In the first case,

$$\theta_0 = 25°C,\ \theta_1 = 70°C$$

$$\theta_2 = 50°C,\quad t = 4 \text{ minutes}$$

$$\int_{70}^{50} \frac{d\theta}{\theta - \theta_0} = -4K$$

$$\log_e\left(\frac{50-25}{70-23}\right) = -4K \qquad ...(i)$$

(2) In the second case

$$\theta_0 = 25°C,\ \theta_1 = 50°C$$

$$\theta_2 = 40°C,\quad t = ?$$

$$\int_{50}^{40} \frac{d\theta}{\theta - \theta_0} = -Kt$$

$$\log_e\left(\frac{40-25}{50-23}\right) = -Kt \qquad ...(ii)$$

Dividing (ii) by (i)

$$\frac{t}{4} = \frac{\log_e\left(\frac{15}{25}\right)}{\log_e\left(\frac{25}{45}\right)}$$

$$t = \frac{4 \times 2.3026}{2.3026}\left[\frac{\log_{10}(0.6)}{\log_{10}(0.55)}\right]$$

$$= 4\left[\frac{\log_{10}(0.6)}{\log_{10}(0.55)}\right]$$

$$= 4\left[\frac{\bar{1}.7782}{\bar{1}.7404}\right] = 4\left[\frac{-2.2218}{-0.2596}\right]$$

$$= \frac{8872}{2596}$$

$$= \mathbf{3.418\ min.}$$

Example 5:

A liquid cools in 6 minutes from 80°C to 60°C. What will be its temperature after the next 10 minutes? Temperature of the surroundings in 30°C. Assume that the Newton's law of cooling is applicable throughout the process.

Solution:

We have $\int_{\theta_1}^{\theta_2} \frac{d\theta}{\theta - \theta_0} = -K\int dt = -Kt$

(1) In the first case,

$$\theta_0 = 30°C,\ \theta_1 = 80°C$$

$$\theta_2 = 60°C, \quad t = 6 \text{ minutes}$$

$$\int_{80}^{60} \frac{d\theta}{\theta - \theta_0} = -6K$$

$$\log_e\left(\frac{60-30}{80-30}\right) = -6K \qquad ...(i)$$

(2) In the second case

$$\theta_0 = 30°C,\ \theta_1 = 60°C$$

$$\theta_2 = x, \qquad t = 10 \text{ min.}$$

$$\int_{60}^{x} \frac{d\theta}{\theta - \theta_0} = -10K$$

$$\log\left(\frac{x - 30}{60 - 30}\right) = -10K$$

Dividing (ii) by (i)

$$\frac{\log\left(\frac{x - 30}{60 - 30}\right)}{\log(0.6)} = \frac{10}{6}$$

$$\log\left(\frac{x - 30}{60 - 30}\right) = \frac{5}{3}\log(0.6)$$

$$\log\left(\frac{x - 30}{30}\right) = \log(0.6)^{5/3}$$

$$\frac{x - 30}{30} = (0.6)^{5/3}$$

$$\mathbf{x = 42.80°C.}$$

Example 6:

A body initially at 80°C cools to 64°C in 5 minutes and to 52°C in 10 minutes. What will be its temperature after 15 minutes and what is the temperature of the surroundings?

Solution:

We have $\int_{\theta_1}^{\theta_2} \frac{d\theta}{\theta - \theta_0} = -K\int dt = -Kt$

(1) In the first case,

$$\theta_1 = 80°C,\ \theta_2 = 64°C$$

$$t = 5 \text{ min.}$$

$$\int_{80}^{64} \frac{d\theta}{\theta - \theta_0} = -5K$$

$$\log\left(\frac{64 - \theta_0}{80 - \theta_0}\right) = -5K \qquad ...(i)$$

(2) In the second case

$$\theta_1 = 64°C, \theta_2 = 52°C$$

$$t = 10 - 5 = 5 \text{ min}$$

$$\int_{64}^{52} \frac{d\theta}{\theta - \theta_0} = -5K$$

$$\log\left(\frac{52 - \theta_0}{64 - \theta_0}\right) = -5K \qquad ...(ii)$$

From equations (i) and (ii)

$$\frac{64 - \theta_0}{80 - \theta_0} = \frac{52 - \theta_0}{64 - \theta_0}$$

$\therefore$ **q_0 = 16°C.**

(iii) Let the temperature after 15 minutes be θ_2

$$\theta_1 = 52°C, \theta_0 = 16°C$$

$$t = 15 - 10 = 5 \text{ min}$$

$$\int_{52}^{\theta_2} \frac{d\theta}{\theta - \theta_0} = -5K$$

$$\log\left(\frac{\theta_2 - 16}{52 - 16}\right) = -5K \qquad ...(iii)$$

Equations (ii) and (iii)

$$\frac{\theta_2 - 16}{52 - 16} = \frac{52 - 16}{64 - 16}$$

q_2 = 43°C

Example 7:

Find the specific heat of a liquid which takes 2 minutes in cooling from 50°C to 40°C in a vessel in which the same volume of water takes

5 minutes in cooling through the same range, of temperature. Mass of water = 100g, mass of liquid = 85g, water equivalent of the vessel = 10g.

Solution:

We have $\frac{MC + w}{t_2} = \frac{m + w}{t_1}$

Here $M = 85g, \quad C = ?,$

$w = 10g$

$t_2 = 2 \text{ minutes} = 120s$

$t_1 = 5 \text{ minutes} = 300s$

$$C = \left[\frac{(m + w)t_2}{Mt_1}\right] = \frac{w}{M}$$

$$C = \left[\frac{(100 + 10)120}{85 \times 300}\right] - \frac{10}{85}$$

C = 0.4 calories/g-K.

Example 8:

Find the value of the universal gas constant R for one gram molecule of a gas.

Solution:

One gram molecule of a gas at N.T.P. occupies 22400 cm^3

$P = 76 \text{ cm of Hg} = 76 \times 13.6 \times 981 \text{ dynes/cm}^2$

$V = 22400 \text{ cm}^3,$

$T = 273K$

$PV = KT$

or $$R = \frac{PV}{T} = \frac{76 \times 13.6 \times 981 \times 22400}{273}$$

$\mathbf{R = 8.31 \times 10^7}$ **ergs/mole-K.**

Note : The value of R is the same for all gases, provided the mass of the gas is one gram molecule.

Example 9:

Calculate the specific heat of air at constant volume, given that specific heat at constant pressure is 0.23, density of air at N.T.P. 1.293 gram/litre and $J = 4.2 \times 10^7$ ergs/cal, $C_v = ?$

Solution:

$C_p = 0.23$, $J = 4.2 \times 10^7$ ergs/cal, $C_p = ?$

Density of air at N.T.P. = 1.293 g/litre

Volume of one gram of air at N.T.P.

$$= \frac{1000}{1.293} \text{ cc}$$

$$PV = rT$$

$$r = \frac{PV}{T}$$

$$= \frac{76 \times 13.6 \times 980 \times 1000}{273 \times 1.293}$$

$$C_p - C_v = r/J$$

$$C_v = C_p - r/J$$

$$= 0.23 - \frac{76 \times 13.6 \times 980 \times 1000}{273 \times 1.293 \times 4.2 \times 10^7}$$

$$C_v = 0.23 = 0.0683$$

$$= \mathbf{0.1617.}$$

Example 10:

Calculate the difference in the two specific heats of one gram of helium, given that the molecular weight of helium = 4 and the gram molecular volume of helium at N.T.P. = 22.4 litres.

Solution:

We have $C_p - C_v = \frac{r}{J}$

Volume of one gram of helium at N.T.P..

$$V = \frac{22400}{4} = 5600 \text{ cm}^3$$

$$PV = rT$$

$$r = \frac{PV}{T}$$

or $$r = \frac{76 \times 13.6 \times 980 \times 5600}{273}$$

$$Cp - C_v = \frac{r}{J}$$

$$r = \frac{76 \times 13.6 \times 980 \times 5600}{273 \times 4.2 \times 10^7}$$

$$= \mathbf{0.4946.}$$

Example 11:

An engine consumes 25 litres of gasoline per hour. The calorific value of gasoline is 6 × 10^6 calories per litre. The output of the engine is 35 kilowatts. Calculate the efficiency of the engine.

Solution:

Total heat produced by gasoline in one hour = $25 \times 6 \times 10^6$ cals

$$\text{Heat produced per second} = \frac{25 \times 6 \times 10^6}{3600} \text{ cal/s}$$

$$\text{Input} = \frac{25 \times 6 \times 10^6 \times 4.2}{3600} \text{ joules/s, i.e., watts}$$

[1 calorie = 4.2 joules]

Output = 35000 watts

$$\text{Efficiency} = \frac{\text{Useful output}}{\text{Input}} = \frac{35000 \times 3600}{25 \times 6 \times 10^6 \; 4.2}$$

$$= 1/5$$

% efficiency = 20%.

EXERCISES

1. Discuss Dulong and Petit's law and explain the variation of the atomic heat of a substance with temperature.
2. How do you explain Dulong and Petit's law according to quantum theory of radiation.
3. Discuss Debye-Einstein theory regarding, atomic heats of solids. Discuss Debye's T^3 law.
4. Deduce the relation between the specific heat of a gas at constant pressure and at constant volume.
5. Describe how the specific heat of a gas at constant pressure is determined accurately.
6. Describe Nernst vacuum calorimeter and indicate briefly how it may be used to determine the specific heat at low temperatures.
7. Describe fully the working of a continuous flow electric calorimeter for the measurement of C_p of a gas.
8. Distinguish between the specific heat of a gas at constant pressure and at constant volume. Describe an accurate method to determine the specific heat of a gas at constant pressure.
9. Describe Jolly's differential steam calorimeter. How will you use this apparatus for determining the specific heat of a gas at constant volume ? Mention the sources of error.
10. Point out the difference between the two principal specific heats of a gas. Describe an accurate method to determine the specific heat of a gas at constant pressure.
11. State and explain Newton's law of cooling. How will you find the specific heat of a liquid by the method of cooling?
12. Describe Nernst vacuum calorimeter for determining the specific heat of a good-conducting solid at low temperatures.
13. Describe the nature of variation of specific heat of solids with temperature. Discuss important regions of this curve.
14. A liquid takes 10 minutes to cool from 90°C to 60°C. How much time will it take to cool from 70°C to 40°C? The temperature of the surroundings is 30°C.

15. A body takes 6 minutes to cool from 80°C to 50°C. How much time will it take to cool from 60°C to 30°C ? The temperature of the surroundings is 20°C.

16. A body cools from 70°C to 50°C in 6 minutes. What will be its temperature after the next 6 minutes? The temperature of the surroundings is 15°C.

17. A liquid cools in 7 minutes from 60°C to 40°C. What will be its temperature after the next 7 minutes ? The temperature of the surroundings is 10°C.

18. A liquid takes 5 minutes to cool from 70°C to 50°C. How much time will it take to cool from 50°C to 35°C? The temperature of the surroundings is 20°C.

19. A body takes 3 minutes to cool from 70°C to 60°C. What will be its temperature after the next 6 minutes? The temperature of the surroundings is 25°C.

20. A liquid takes 4 minutes to cool from 65°C to 50°C. What will be its temperature after the next 10 minutes? The temperature of the surroundings is 35°C.

21. A petrol engine consumes 25 kg of petrol per hour. The calorific value of petrol is 11.4×10^6 cals per kg. The power of the engine is 99-75 kilowatts. Calculate the efficiency of the engine.

22. A Nernst calorimeter of lead weighing 396.3 grams and surrounded by a bath at temperature 61 K, on being heated electrically for four minutes showed a rise in temperature of 1.219 K under condition of no heat leakage. Calculate the atomic heat of lead at 61 K, if the potential drop and average current during the interval of heating were 1.586 volts and 0.1444 ampere respectively. (Atomic weight of lead = 206.4 and J = 4.18 joules/calorie.)

23. A copper calorimeter of mass 100 g containing 150 cm^3 of a liquid of specific heat 0.6 and specific gravity 1.2, is found to cool at the rate of 2°C per minute when its temperature is 50°C above that of the surroundings. If the liquid is emptied out and 150 cm^3 of a second liquid of specific heat 0.4 and specific gravity 0.9 are substituted, what will be its rate of cooling, when the temperature is 40°C above that of its surroundings? (Sp. heat of copper 0.1.).

24. Describe the continuous flow calorimeter. What are the advantages of this method? How do you determine C_p of a gas by using this method?

25. Describe fully the working of the continuous now calorimeter of 'Callendar and Barnes'. What are the advantages of this method ? What information has it given regarding the specific heat of water?

26. Derive an expression for the difference in the two specific heats of a gas.

27. Describe and explain Joly's steam calorimeter method for finding the specific heat of a gas at constant volume.

28. Point out the difference between the two principal specific heats of a gas and show that for an ideal gas

$$C_p - C_v = R.$$

29. Describe the bomb calorimeter method to find the calorific value of a fuel.

30. Indicate the manner in which the specific heat of a solid varies with temperature and outline the theory which explains this variation.

31. Discuss Newton's law of cooling. How will you determine the specific heat of a liquid by Newton's law of cooling?

32. Describe the Regnault's method to find the specific heat of a gas at constant pressure.

33. Discuss the working of a bell calorimeter in determining the calorific value of a fuel.

4

Change of State

4.1 INTRODUCTION

A substance can exist in three states *viz.*, solid, liquid and gas. The particular state of a substance depends on its temperature. According to the kinetic theory, the molecules of a substance in the solid state have less degrees of freedom than a substance in the liquid or gaseous state. The molecules are more free to move in the gaseous state. The change of state can be brought about by supplying or withdrawing heat from the substance. Ice is the solid state of water. By supplying heat to ice, it can be changed into water. Similarly by supplying heat to water, it can be converted into the gaseous state (steam). The reverse processes occur when heat is withdrawn from steam. This is true for all substances. Even permanent gases like oxygen, nitrogen, hydrogen etc., can be liquefied at low temperatures.

For a given substance, the change of state takes place at a fixed temperature and at a given pressure and vice-versa. When the substance changes from the solid to the liquid state, the heat supplied at constant temperature (called melting point) is used in overcoming the forces of intermolecular attraction. The mean molecular distance increases and the molecules are more free to move. Similarly, when the substance changes from the liquid to the gaseous state at a fixed temperature (called the boiling point) at standard atmospheric pressure, the heat supplied is used in increasing the mean molecular distance. The molecules become free to move about in the whole space available to them.

4.2 LATENT HEAT OF FUSION

Take small pieces of ice in a beaker. Fix a thermometer to note the temperature of ice in the beaker. Heat the beaker slowly. Ice melts but the thermometer does not show any rise in temperature. When the whole ice has melted the temperature of water rises and it is indicated by the thermometer. During the processes of conversion from ice to water (change of state from solid to liquid) the heat supplied is used to change the state of ice from solid state to liquid state. This heat is called latent heat. Latent means hidden *i.e.*, which is not indicated by the thermometer.

It has been found that one gram of ice takes 80 calories of heat to get itself converted to water. This heat is called latent heat of fusion of ice. Its unit is cals/gram.

Latent heat of fusion of a substance is defined as the amount of heat required to change the state of one gram of a substance from solid to liquid without any change in its temperature.

The melting point of ice is 0^0C at a pressure of 76 cm of Hg.

4.3 LAWS OF FUSION

1. Every substance changes its state from solid to liquid at a particular temperature (under normal pressure) called the melting point.

2. As long as the change of state takes place, there is no change in temperature.

3. One gram of every substance requires a definite quantity of heat for change of state from solid to liquid and it is called the latent heat of fusion of that substance. It is different for different substances.

4. Some substances show increase in volume on melting *e.g.*, wax, ghee etc. while some other substances show decrease in volume on melting *e.g.*, ice.

5. The melting point of those substances which decrease in volume on melting, is lowered with increase in pressure.

6. The melting point of those substances which increase in volume on melting, is increased with increase in pressure.

Notes:

(i) Ice decreases in volume on melting and its melting point is lowered with increase in pressure.

(ii) The melting point or solidification point is the same for a substance.

4.4 PRACTICAL APPLICATIONS

1. Aquatic animals live underneath water in ponds and lakes in winter. Ice is formed on the surface of the pond, and it remains on the surface due to its increase in volume and hence lesser density.
2. Rocks in the mountains break in frosty weather. Water, which may be locked up in the pores of the rock, may freeze into ice and there is increase in volume. This produces cracks in the rocks.
3. Fertility of the soil increases due .to the splitting of the soil in frosty weather. Water freezes into ice and there is increase in volume. This increase in volume of ice disintegrates the soil.
4. Alloys which expand on solidification are used for better casting. The alloy in the liquid form is put into the cast. When it solidifies, there is increase in volume and a sharp casting is obtained. Type metal expands on solidification.

4.5 EFFECT OF PRESSURE ON FREEZING POINT OF ICE—REVELATION

Melting point of ice is lowered with increase in pressure. Take a slab of ice. Take a wire and fix two weights (5 kg each) at its two ends. Put the wire over the slab as shown in (Fig. 4.1).

The wire passes through the ice slab. Just below the wire, ice melts at a lower temperature due to increase in pressure. When the wire has passed, the water above the wire freezes again. Thus the wire passes through the slab and the slab does not split. This phenomenon of refreezing is called *Regelation*.

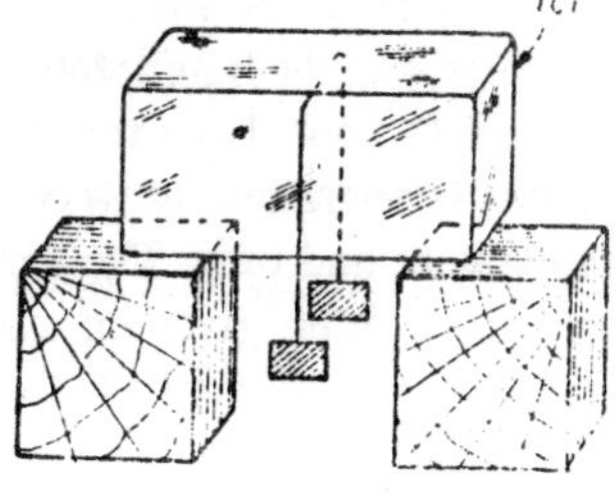

Fig. 4.1

In a similar way, it is possible to freeze two-pieces of ice by applying pressure and then releasing them.

Examples

1. Water pipes burst in winter in cold countries due to solidification of water into ice. When water is converted into ice there is increase in volume and the pipes are likely to burst.
2. When wheels of a cart pass over snow, it melts due to increase in pressure exerted by the wheels. When the wheel passes over, the water formed on the wheel solidifies due to regelation. Due to this reason wheels are covered with snow.
3. Skating is possible on snow due to the formation of a thin layer of water formed below the skates. The water is formed due to the increase of pressure and it acts as a lubricant.

4.6 IMPURITIES LOWER FREEZING POINT

When salt is mixed, with ice, some ice melts taking heat from the salt. The temperature of the mixture decreases. Further, the salt dissolves in the water formed and takes latent heat from the mixture. The temperature of the mixture is decreased further. In this way, with a freezing mixture of salt and ice in the ratio 1 : 3, temperatures as low as –13°C can be obtained. A freezing mixture consists of powdered ice, common salt and ammonium nitrate.

Similarly, when sugar is dissolved in water, the water is cooled.

4.7 DETERMINATION OF MELTING POINT OF WAX

Take sufficient quantity of wax in a test tube and fix a cork along with a thermometer. Adjust the test tube in a beaker containing water (Fig. 4.2). Heat the water to its boiling point. The wax in the test tube melts and is in the liquid state. Extinguish the burner and allow the water and wax to cool. Note the temperature of wax after every one minute till the temperature of wax is about 30°C. Draw a graph between temperature and time. The temperature corresponding to the horizontal line in the graph gives the melting point of wax. At this temperature, liquid wax is converted into the solid wax without change of temperature (Fig. 4.3).

Plot a graph between temperature along the Y-axis and time in minutes along the X-axis.

The temperature corresponding to the line BC gives the melting point of wax. The portion AB shows the liquid state of wax and below C the graph represents the solid state.

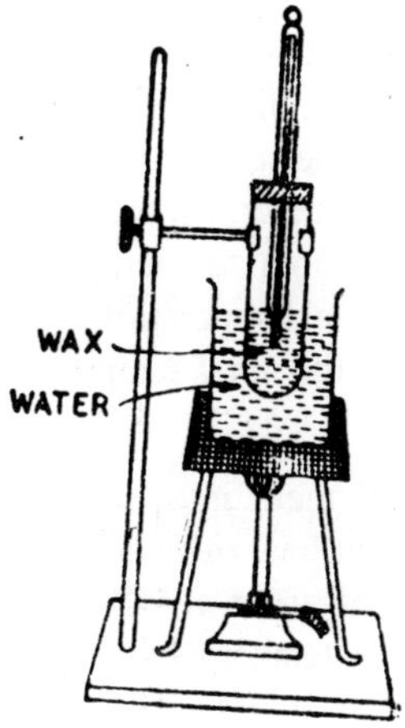

Fig. 4.2

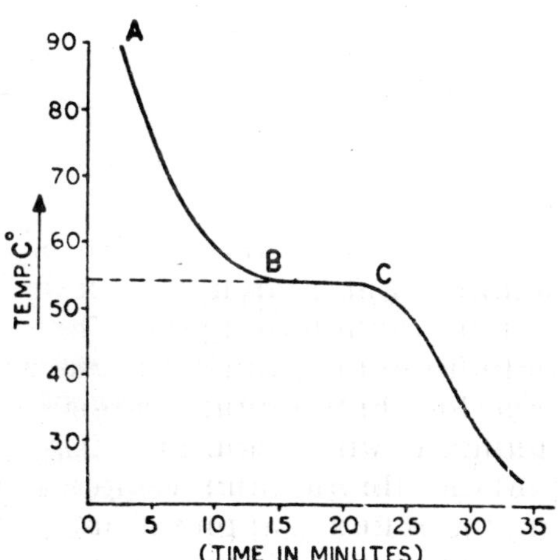

Fig. 4.3

4.8 DETERMINATION OF LATENT HEAT OF FUSION OF ICE

Take a calorimeter. Weigh the calorimeter along with the stirrer. Weigh the calorimeter with stirrer and the lid. Heat some water in a beaker to a temperature of about 10°C higher than the room temperature. Fill the calorimeter to about half, with hot water. Weigh the calorimeter, hot water, stirrer and lid. Place it in a wooden box. Note the temperature of water in the calorimeter. It should be about 4 to 5°C higher than the room temperature. Take small pieces of ice in a blotting paper. Add dry pieces of ice gradually into the calorimeter and stir. Continue adding dry pieces of ice in to the calorimeter, till the temperature of water is about 4 to 5°C lower than the room temperature. (This is done to avoid radiation error).' Note the final temperature and weigh the whole contents. See that before recording the final temperature the whole of ice has melted.

When one gram of ice melts into water at 0°C, the amount of heat absorbed is called the latent heat of fusion of ice. Suppose M grams of ice is added to m grams of water at t°C in a calorimeter whose water equivalent is W. The final temperature reached = T°C.

Heat lost by water and calorimeter = (m + W)(t – T)

Heat gained by ice = ML + MT

Heat gained = Heat lost

$$ML + MT = (m + W)(t - T)$$

$$L = \frac{(M + w)(T - t_1)}{m} - (t_2 - T.$$

Hence, latent heat of fusion of ice can be determined.

4.9 HEAT ABSORBED IN SOLUTION

When a solution is formed by dissolving a solute (solid substance) in a solvent (liquid) the following possible changes take place.

1. The solute gets dissolved in the solvent and takes the necessary latent heat from the solvent. The temperature of the solution decreases. However, in some rarer cases there is absorption of heat.
2. When a solute is dissolved in a solvent, a chemical reaction may, take place. This reaction is generally accompanied by evolution of heat.

3. When an electrolyte is dissolved in water *e.g.*, $CuSO_4$ dissolved in water, ionic dissociation takes place. In this process heat is absorbed.

However, depending on the nature of the substance *i.e.*, solute and the solvent, the total effect may be absorption or evolution of heat.

4.10 VAPORIZATION AND CONDENSATION

The conversion of a substance from the liquid to the vapour state is called vaporization. Boiling is the phenomenon of vaporization accompanied by ebullition *i.e.*, violent ebullition of bubbles. However vaporization can take place without boiling. The water kept on an open plate in summer evaporates and thcrc is a change of state. But there is no boiling. Conversion of- water to steam is vaporization. The conversion of the substance from the vapour state to the liquid state is called condensation. Conversion of steam into water is called condensation. Rain is caused due to the condensation of water vapour in the clouds.

4.11 LAWS OF EBULLITION OR BOILING

1. Every liquid changes its state from liquid to its vapour at a particular temperature (under normal pressure) called the boiling point.
2. As long as the change of state takes place, there is no change in temperature.
3. One gram of every liquid requires a definite quantity of heat for change of state from liquid to vapour and it is called the latent heat of vaporization of that liquid. Latent heat is different for different substances.
4. All liquids show increase in volume on vaporization.
5. The boiling point of a liquid increases with increase in pressure of a liquid.
6. The liquid can boil at a lower temperature under reduced pressure.

4.12 CHANGE IN BOILING POINT WITH PRESSURE

The effect of pressure on the boiling point (saturation temperature) of a liquid can be studied with the help of the apparatus shown in (Fig. 4.4). The temperature at which the change of state from liquid to vapour takes place is called saturation temperature. The saturation temperature increases with increase in pressure. If the pressure is 1 atmosphere, the saturation temperature is called the boiling point. The vessel A contains the liquid (say water). It is connected to the flask B through a tube C surrounded by a water condenser. To measure the pressure of air over the surface of the liquid in A, a manometer is used.

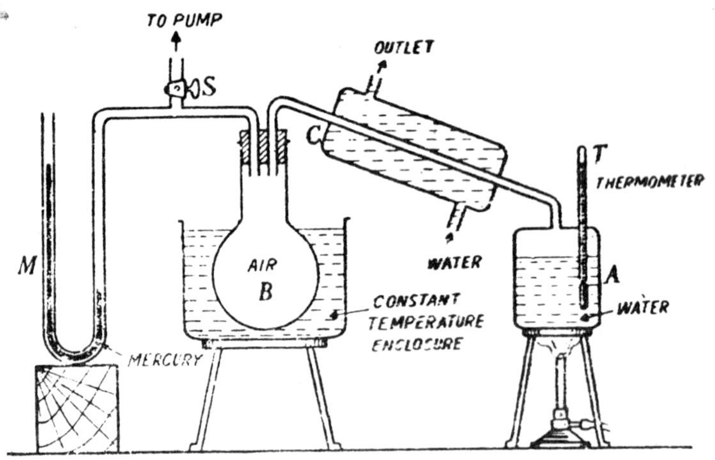

Fig. 4.4

The pressure of air inside A and B can be changed with the help of the compression pump or evacuation pump.

The stopcock S is opened. With the help of the compression pump, air is compressed into B. The stopcock is closed. Pressure of air in A and B is shown by the manometer. It is more than the atmospheric pressure.

The liquid is heated and the temperature at which it boils is recorded by the thermometer. At the boiling point, the temperature of the liquid remains constant. The condenser through which cold water is circulated prevents steam to enter the vessel B. The steam is condensed and condensed water flows back to A.

In this way at various pressures, the corresponding boiling points of the liquid can be determined. For low pressure, the vacuum pump is used. It is found that the boiling point of a liquid increases with increase in pressure and decreases with decrease in pressure.

The boiling point of water at 76 cm of Hg pressure is 100°C. The boiling point of water at double the normal pressure (*i.e.*, 152cm of Hg) is about 128°C.

Applications

Papin's Digester and Pressure cookers are based on the principle that the boiling point of a liquid increases with increase in pressure.

The pressure inside the pressure cooker is much higher than the atmospheric pressure. The steam produced inside the cooker is not allowed to escape and the pressure over the surface of water increases. It means water inside the vessel will be at a temperature higher than the normal boiling point. Vegetables cooked with a pressure cooker are prepared in a short time.

When the pressure increases beyond the desired pressure, the weight placed on the lid lifts up a little and allows the steam to escape slowly and maintains constant pressure inside it. For safety purposes, a safety valve is also provided on the lid. The safety valve on a pressure cooker is a fusible plug. In case, the weight fails to lift, the pressure inside the cooker may increase enormously and may lead to explosion. To prevent this, a fusible plug is provided. It has a small stem made of a material of low melting point. When the pressure inside increases, the saturation temperature also increases and the plug melts. Due to, this the steam escapes and the pressure inside the cooker gets lowered.

When a pressure cooker is kept on a stove, on filling it with water to the normal level, vaporization takes place. The water starts evaporating at temperatures lower than the boiling point temperature because of the pressure of air and hence the actual pressure of evaporation is the difference between the atmospheric pressure and the partial pressure of air. As the temperature increases, the rate of evaporation increases and the air is expelled. Gradually, due to the expulsion of air, the vapour pressure of water increases and hence the temperature. After all the air is expelled, boiling takes place. The steam starts rushing out through the opening.

The weight is now placed on the opening. Further, steam that is produced increases the pressure and hence the boiling point is also raised. As soon as the upward force due to the steam pressure is sufficiently high, the weight lifts and allows the steam to escape. This maintains constant pressure and hence constant temperature. If the weight is placed before the expulsion of steam, the partial pressure will not be as high as it ought to be and hence the temperature is not high.

4.13 FRANKLIN'S EXPERIMENT

Take a flask containing water. Heat it till the water begins to boil. When the water is boiling, the air escapes to the atmosphere. After some time, close the mouth of the flask with an air tight rubber cork. Invert the flask and fix it on the retort stand (Fig. 4.5). Pour cold water over the flask. Due to the condensation of steam, the pressure over the surface of water decreases. Water inside the vessel begins to boil. This experiment shows that water boils at a lower temperature under reduced pressure.

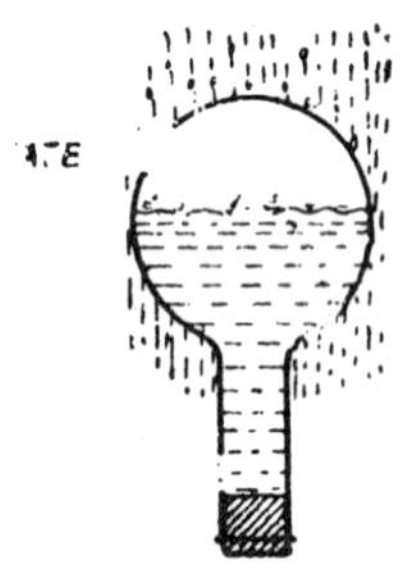

Fig. 4.5

In the laboratory, while preparing steam, the temperature of steam is usually less than 100°C because the atmospheric pressure is less than 76 cm of Hg.

4.14 LATENT HEAT OF VAPORIZATION

Latent heat of vaporization of a liquid is the amount of heat required to convert one gram of a liquid into vapour without any change in temperature. For water, latent heat of vaporization is 537 cals/gram at a pressure of 76 cm of Hg. Latent heat of vaporization is different for different substances.

Determination of Latent Heat of Steam

Take a calorimeter. Weigh the calorimeter along with the stirrer. Weigh the calorimeter along with the stirrer and the lid. Pill the calorimeter to about 2/3 with water and weigh along with the stirrer and the lid. Heat the water in the boiler to produce steam. The boiler is fitted with a steam trap as shown in Fig. 4.6. The whole of the delivery tube is covered with wool. When steam starts coming out of the end of the tube, note the temperature of Steam. Note the temperature of water in the calorimeter. Now clean the end of the tube with a blotting paper to remove water drops. Insert the tube quickly into the calorimeter and pass steam. When the temperature of water is raised through about 10°C, take out the tube carefully. Continue stirring. Note the final temperature. Remove the thermometer carefully and weigh the calorimeter. The calorimeter should be covered with a lid and no water drops should be sticking to the thermometer.

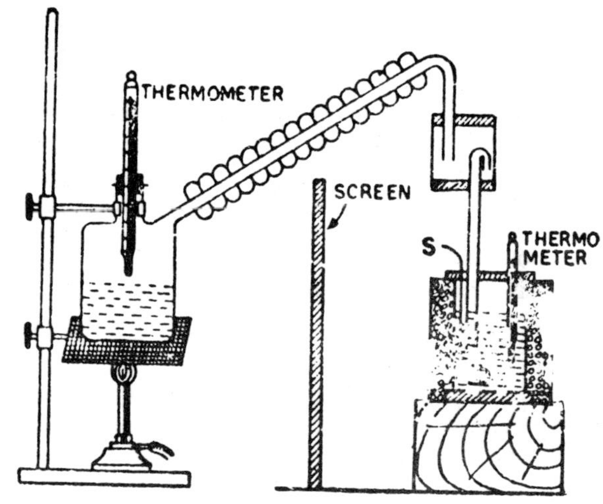

Fig. 4.6

Suppose, m grams of steam is passed into M grams of water at (t°C in a calorimeter of water equivalent w. The temperature of steam is t°C and final temperature is T°C.

Heat gained by water and calorimeter = $(M + w)(T - t_1)$

$$\text{Heat lost by steam} = mL + m(t_2 - T)$$

$$\text{Heat gained} = \text{Heat lost}$$

$$(M + w)(T - t_1) = mL + m(t_2 - T)$$

$$L = \frac{(M+w)(T-t_1)}{m} - (t_2 - T$$

Hence, latent heat of steam can be determined.

4.15 BUNSEN'S ICE CALORIMETER

Bunsen's ice calorimeter is used to find the specific heat of a solid or a liquid available in a small quantity. It is based on the principle that when ice melts there is decrease in volume. When one gram of ice melts, the decrease in volume is 0.09 cc.

The apparatus consists of a metal calorimeter C fitted inside a bulb B. The bulb B is connected to the cup A through a glass tube. The cup A is fitted with an air tight lid. A capillary tube EF is fixed whose one end dips inside mercury in the cup A The horizontal portion of the capillary tube is fitted on a graduated scale. The upper portion of the bulb B contains water at 0°C. The lower portion of B, the connecting tubes and the cup A contain mercury. By adjusting the position of the screw T the position of the mercury meniscus in the capillary tube EF can be suitably adjusted. Th whole apparatus is enclosed in a bath of melting ice.

Working

Ether is poured in the calorimeter C. It evaporates and takes heat for evaporation from the surrounding water in B. As water in B is already at 0°C, a small quantity of water surrounding C freezes into ice.

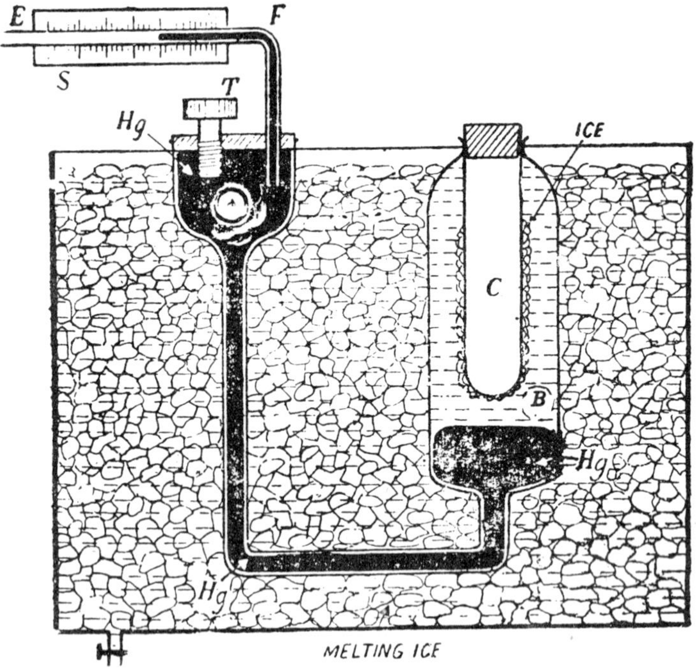

Fig. 4.7

The substance whose specific heat is to be determined is heated to a desired temperature (θ°C). The position of mercury in the capillary EF is noted on the scale.

The substance is transferred quickly into the calorimeter C. The calorimeter is closed with the rubber cork. The substance loses heat to the surrounding ice and cool to 0°C. A small quantity of ice melts and there is decrease in volume. Consequently the mercury in the capillary tube EF recedes. Suppose the length through which mercury has moved = l cm and area of cross-section of the capillary tube = a sq cm.

∴ Decrease in volume = v = al

When decrease in volume is 0.09 cc the amount of ice that melts

= 1 gram

When decrease in volume is vcc the amount of ice that melts

$$= m = \frac{v}{0.09} \text{ grams}$$

Latent heat of ice = 80 cal/g

Heat gained by ice $= 80\,m = \dfrac{80\,v}{0.09}$ cal

Heat lost by the substance

= MSB

Heat lost = Heat gained

$$\therefore \quad MS\,\theta = \frac{80v}{0.09}$$

$$\text{or} \quad S = \frac{80v}{0.09\,M\theta} \text{ cal/g}-°C$$

Advantages

(i) This method is accurate and sensitive.

(ii) It is suitable to find the specific heat of a substance available in small quantity.

(iii) Water equivalent of the calorimeter does not come into the calculation as the calorimeter remains at 0°C throughout.

(iv) There is no need to note the final temperature.

(v) There is no Ion of heat by radiation or conduction.

4.16 COOLING EFFECT DUE TO VAPORIZATION

If a person pours a little spirit on his hand, he feels a cooling effect. The spirit evaporates and takes latent heat of vaporization from the spirit left behind and the hand. Due to this reason one feels a cooling effect

Thus, whenever evaporation of a liquid takes place, the remaining liquid shows a fall in temperature.

The molecules of a liquid are always in a state of random motion. They collide with each other and they may or may not change their respective kinetic energies. Due to the collision a certain molecule may acquire more energy than the energy with which it is held. As soon as this happens the molecule flies off, *i.e.*, it leaves the liquid surface. This is called evaporation. If no external heat is supplied it is obvious that the mean kinetic energy of the molecules decreases and the temperature of the liquid falls. This explains cooling caused by evaporation. If the moving vapour molecules come near to each other at sufficiently low velocities, they experience a greater magnitude of force of attraction. The molecules coalesce and this results in condensation. Rate of evaporation increases with increase in temperature of the liquid and decrease in pressure over the liquid surface.

4.17 CRYOPHORUS

A cryophorus consists of two glass bulbs A and B connected through a glass tubing 0 (Fig. 4.8). The bulb A contains water.

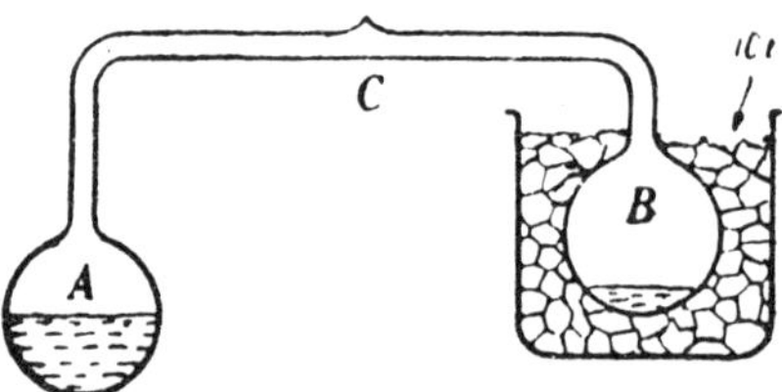

Fig. 4.8

Air in B and the tube is removed with an evacuation pump and the tube is scaled. The space inside B and the tube contains water vapour only. The vessel B is surrounded by freezing mixtures. Water vapour in B condenses and the pressure over the surface of water in A falls.

Evaporation of water takes place. The process continues at a rapid rate and the temperature of water in A decreases. After some time water in A begins to freeze.

To prevent any heat from the surroundings to flow into A, it is covered with a non-conducting material like cotton or flannel.

4.18 EXAMPLES OF COOLING CAVED BY EVAPORATION

1. In hot weather, we perspire. Water from the body comes out of the fine pores of the skin. Water evaporates and takes heat of vaporization from the body. Due to this reason we feel a cooling sensation.
2. A person wearing wet clothes feels chill. Water evaporates taking the latent heat of vaporization from the body and the person feels chill.
3. Water in an earthen pot remains cool in summer. Water comes out of the pores of the vessel and evaporates. During evaporation, the latent heat is taken from the remaining water. Therefore water remains cool in an earthen vessel.
4. Streets, floors, gardens, lawns etc. are watered in summer. Water evaporates taking the latent heat from them and keeps them cool.
5. Dogs keep their tongues usually out in summer. Water evaporates from the tongue and keeps it cool.
6. *Khas-khas* screens are commonly used in summer. Water is poured on them. Water evaporates and takes heat from them and the air in the room. Due to this reason the room is kept cool.
7. During high fever, a wet cloth soaked in 'Eau de cologne' or ice cold water is kept on the forehead. The liquid evaporates rapidly and takes heat from the head and the body.
8. Hot liquids are poured in a plate to increase the surface area. The rate of evaporation increases and the liquid cools quickly.
9. When a wet muslin piece is wrapped around the bulb of a thermometer, the thermometer records fall of temperature. Water evaporates and takes latent heat from mercury in the bulb.

10. In a packed .hall in summer a person feels uneasy. Perspiration from his body does not evaporate. As soon as he comes out of the hall he feels a sudden cooling sensation. Perspiration evaporates rapidly and takes heat from his body.

4.19 REFRIGERATION

The second law of thermodynamics as given by Clausius gives the principle of working of a refrigerator and ice plants.

"It is impossible to make heat flow from a body at a lower temperature to a body at a higher temperature without doing external work on the working substance."

This part is applicable in the case of ice plants and refrigerators. Heat itself cannot flow from a body at a lower temperature to a body at a higher temperature. But, it is possible, if some external work is done on the working substance. Take the case of an ammonia ice plant. Ammonia is the working substance. Liquid ammonia at low pressure takes heat from the brine solution in the brine tank and is converted into low pressure vapour. External work is done to compress the ammonia vapour to high pressure. This ammonia at high pressure is passed through coils over which water at room temperature is poured. Ammonia vapour gives heat to water at room temperature and gets itself converted into liquid again. This high pressure liquid ammonia is throttled to low pressure liquid ammonia. In the whole process ammonia (the working substance) takes heat from brine solution (at a lower temperature) and gives heat to water at room temperature (at a higher temperature). This is possible only due to the external work done on ammonia by the piston in compressing it. The only work of electricity in the ammonia ice plant is to move the piston to do external work on ammonia. If the external work is not done, no ice plant or refrigerator will work. Hence, it is possible to make heat flow from a body at a lower temperature to a body at a higher temperature by doing external work on the working substance.

4.20 AMMONIA ICE PLANT

Ammonia ice plant is commonly used to manufacture ice. It is based on the principle that evaporation causes cooling. Moreover in this plant, heat flows from a body at a lower temperature to a body at a higher temperature by doing external work on the working substance. The schematic sketch of an ammonia ice plant is shown in (Fig. 4.9).

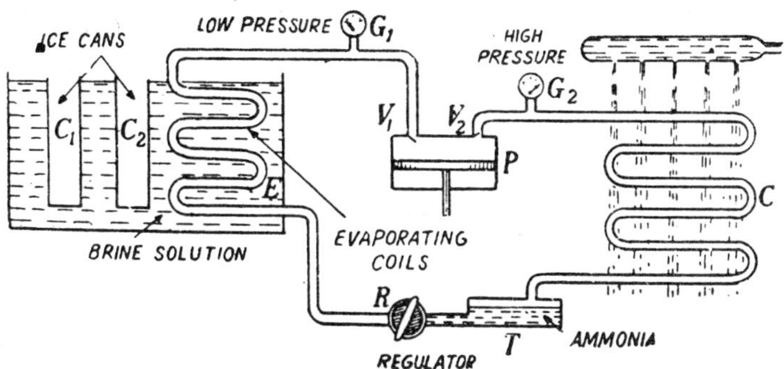

Fig. 4.9

1. T is a reservoir containing ammonia liquid.
2. R is a regulator called the expansion valve or throttle valve. High pressure liquid expands towards the low pressure side and with the fall in pressure there is a fall in saturation temperature. Liquid ammonia at low temperature and pressure entering the coils evaporates and takes heat from the surrounding brine solution.
3. The brine tank contains ice cans C_1 and C_2 containing pure water.
4. The compression pump P is worked with the help of an electric motor. It has two valves V_1 and V_2. When the piston is moved outwards, valve V_1 opens and V_2 is closed. Ammonia vapour enters the cylinder. When the piston is moved inwards V_1, closes and V_2 opens. Compressed ammonia vapour is passed through the condensing coils C.
5. G_1 is the low pressure gauge and G_2 is the high pressure gauge. When the pump works, the rate of evaporation of ammonia liquid in the evaporating coils increases and the temperature of the brine solution gradually falls.
6. The compressed ammonia vapours passing the condensing coils give heat to water falling over the coils. Ammonia vapours get condensed and liquid ammonia is formed. The liquid ammonia enters the reservoir.

The cycle of operations is repeated and the temperature of brine solution in the tank decreases to –8°C. Water in the cans freeze to ice.

(i) If the pump stops working, then the cyclic process does not take place and the temperature of the brine tank does not fall.

(ii) The function of electricity is to move the piston to and from with the help of the electric motor and to compress the vapours.

(iii) Ammonia is the working substance. It takes heat from the brine solution (at low temperature) and transfers heat to the water falling on the condensing coils at the room temperature. It means heat is transferred from a body at a lower temperature to a body at a higher temperature. Ordinarily it is not possible. But by doing external work (with the help of the pump) on the working substance this transfer of heat is possible and this is in accordance with the Second Law of Thermodynamics.

4.21 SOLID CARBON DIOXIDE—(DRY ICE)

A cylinder containing liquid CO_2 is taken. A muslin bag is tied to the nozzle. The cylinder is placed in a slanting position (Fig. 4.10). The valve V is opened and the liquid CO_2 enters the muslin bag. The liquid CO_2 evaporates rapidly and takes the latent heat of vaporization from the remaining liquid in the bag. Liquid CO_2 solidifies and solid CO_2 is formed in the muslin bag.

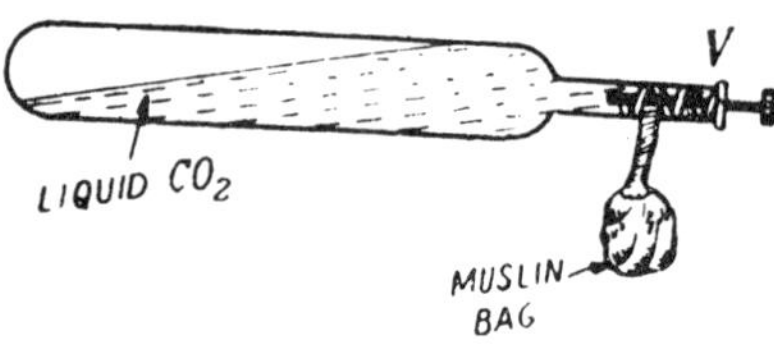

Fig. 410

The temperature of solid CO_2 is –78°C and it is also called *dry ice*. It is commonly used by doctors for dressing the wounds. It serves as refrigerant when mixed with ether. Solid CO_2 does not melt when exposed to air. It directly evaporates and forms its vapours.

4.22 GAS AND VAPOUR

Gas and vapour states are the two distant stages of the same continuous phenomenon. These two states have distinct boundary which is governed by a particular temperature called the critical temperature. Above the critical temperature the substance is in the gaseous state and below the critical temperature it is in the vapour state.

A gas cannot be liquefied by mere application of pressure, howsoever high the pressure may be. A vapour can be liquefied by applying pressure. Therefore to liquefy a gas, first it has to be cooled to a temperature lower than its critical temperature. It means it is converted from a gaseous state to the vapour state. Then a high pressure is applied and the vapours get liquefied.

Critical temperature of a gas is that temperature above which the gas cannot be liquefied by mere application of pressure, howsoever large the pressure may be.

Table 4.1 : Critical Temperature.

Substance	Critical Temperature °C
Water	374
Ammonia	132
Carbon dioxide	31
Oxygen	–119
Hydrogen	–240
Helium	–268

4.23 SATURATED AND UNSATURATED VAPOURS

Take three clean barometer tubes A, B and C as shown in Fig. 4.11. The level of mercury in each tube is up to the same height. The space above mercury in all the tubes is Torricellian vacuum.

With the help of a bent pipette introduce water into the tube B. Water rises to the top and evaporates. Level of mercury in the tube B falls. The space above mercury in B contains water vapours. There is no water left on the surface of mercury. These vapours are unsaturated and the pressure exerted by the vapours is h.

Similarly, continue, introducing water in the tube C. The level of mercury gradually falls. A stage is reached when no more water introduced m the tube C evaporators (Fig. 4.11). It means the space above mercury in C is completely saturated with water vapours. It cannot contain more vapours. The difference in the levels of mercury in A and C is H and it represents the saturated vapour pressure of water at the room temperature.

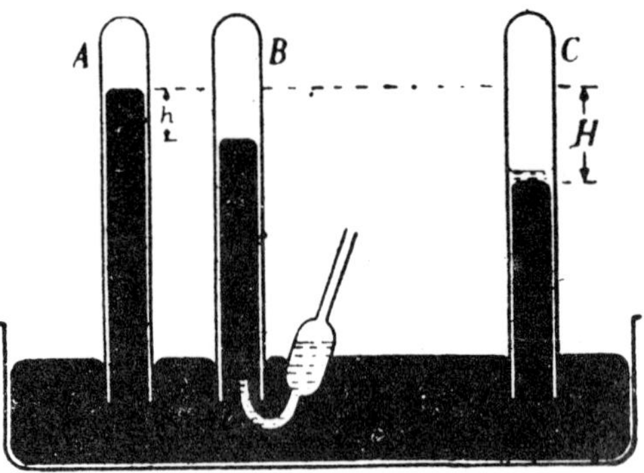

Fig. 4.11

A space is said to be saturated with water vapours if it cannot contain more vapours. It is unsaturated if it can contain more vapours.

1. The saturated vapour pressure of a liquid depends on its temperature. The saturated vapour pressure increases with rise in temperature and decreases with fall in temperature.
2. The saturated vapour pressure depends on the nature of the substance.
3. The saturated vapour pressure does not depend on the volume occupied by the vapours.
4. The saturated vapour of a liquid is independent of the presence of the vapours of other liquids present. The vapours should not have any chemical action:
5. The total pressure exerted by the vapours of all substances is equal to the sum of the pressures exerted by the vapours of individual substances,

$$P = P_1 + P_2 + P_3 + \ldots$$

6. The saturated vapours do not obey the gas laws whereas unsaturated vapours obey the gas laws.

In Fig. 4.12, the curve AB shows that with decrease in volume the pressure increases. From A to B the vapour is unsaturated and obeys Boyle's :aw.

At B the vapour is saturated and exerts maximum pressure. With decrease in volume beyond, B, the pressure is constant. From B to C it represents the change from the saturated vapour to the liquid state.

If the temperature of a saturated vapour is gradually increased the pressure increases. From A to B the vapour is saturated

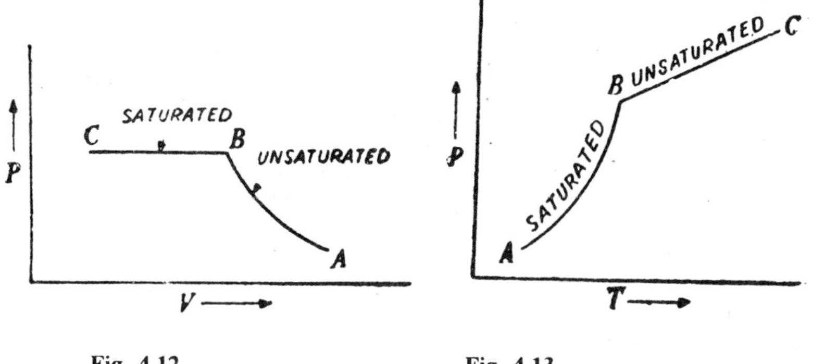

Fig. 4.12 **Fig. 4.13**

(Fig. 4.13). Beyond B if the temperature is increased, the vapour is unsaturated and pressure increases with temperature. In the region BC, the pressure of the unsaturated vapour is directly proportional to its absolute temperature.

4.24 VAPOUR PRESSURE OF LIQUIDS

The vapour pressure of a liquid is dependent only on its temperature. The vapour pressure, in general, increases with rise in temperature. The normal boiling point is that temperature at which the vapour pressure of the substance is equal to the normal pressure *i.e.*, 76 cm of Hg. If the atmospheric pressure is lower or higher than the normal pressure, the liquid will boil at a temperature lower or higher than the normal boiling point. Usually the atmospheric pressure in Delhi is less than 76 cm of Hg and water boils at a temperature lower than 100°C. For the measurement of vapour pressure the apparatus shown in Fig. 4.14 is used.

The bulb B contains the vapour in contact with its liquid. M is a manometer limb and B is a reservoir containing mercury. At high temperatures, the stop-cock S is kept open and the tube is filled with a gas at high pressure. Adjust the reservoir R such that the liquid appears in the tube inside the oven. If the pressure of the gas in the limb A is

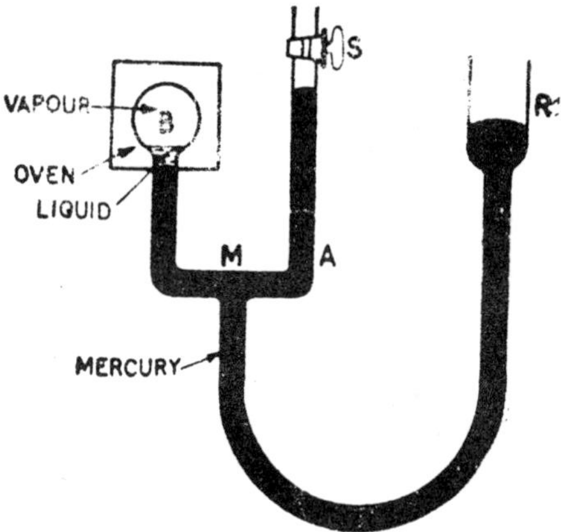

Fig. 4.14

P_0 and the difference in levels of mercury in the manometer limbs is h, then vapour pressure of the liquid at the temperature T is given by

$$P = P_0 + h \qquad ...(i)$$

In this way the vapour pressure of the liquid at any temperature can be determined.

For low vapour pressure measurement (*i.e.*, at low temperatures) the reservoir S is raised so that the level of mercury in the limb A is above the stop-cock S. The stop-cock S is closed. The vessel B is surrounded by the low temperature bath and the reservoir R is adjusted so that the liquid in the tube is within the bath. The difference in the levels of mercury in the manometer directly measures the vapour pressure. In this case $P_0 = 0$

$$\therefore \qquad P = h \qquad ...(ii)$$

As the temperature of the liquid is lowered, a state is reached when the liquid begins to solidify. The temperature and pressure correspond to the triple point. At the triple point the substance exists as a solid, liquid and vapour simultaneously.

4.25 TRIPLE POINT

(i) The melting point of ice decreases with increase in pressure. It means that ice melts at a temperature lower than 0°C at a pressure higher than the normal pressure. The curve AB

(Fig. 4.15) represents the relation between pressure and temperature. The curve AB is called the ice line. The substance will exist in the solid state (ice) to the left of the curve AB and in the liquid state to the right of the curve AB. The curve AB represents the equilibrium between the liquid and the solid states.

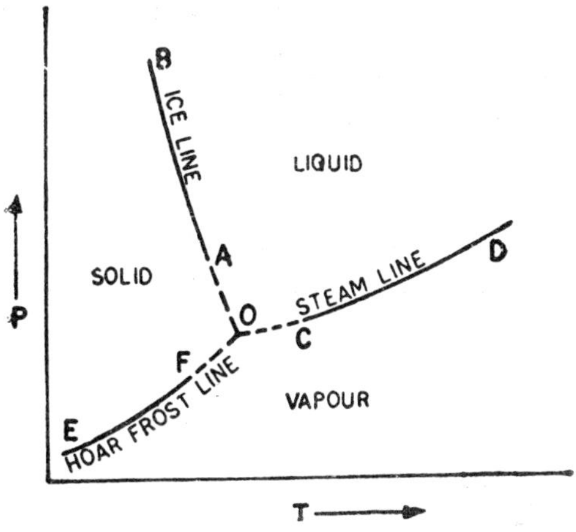

Fig. 4.15

(ii) The boiling point of water increases with increase in pressure. The curve CD represents the relation between pressure and temperature and it is called the steam line. Above the curve CD, the substance is in the liquid state and below the curve CD it is in the vapour state. The curve CD represents the equilibrium between the liquid and the vapour states.

(iii) The curve Ef represents equilibrium between the solid and the vapour states of a substance and is called the floor frost line. Above the curve EF, the substance is in the solid state and below the curve EF it is in the vapour state.

These three curves should meet at a point O as shown in Fig. 4.15. Suppose these curves do not meet at a point as shown in Fig. 4.16 and they enclose an area ACF. It means, according to the ice line, the substance must be only in the solid state in the shaded portion. According to the steam line, the substance must be only in the liquid state and according to the hoar frost line the substance must be only in the vapour

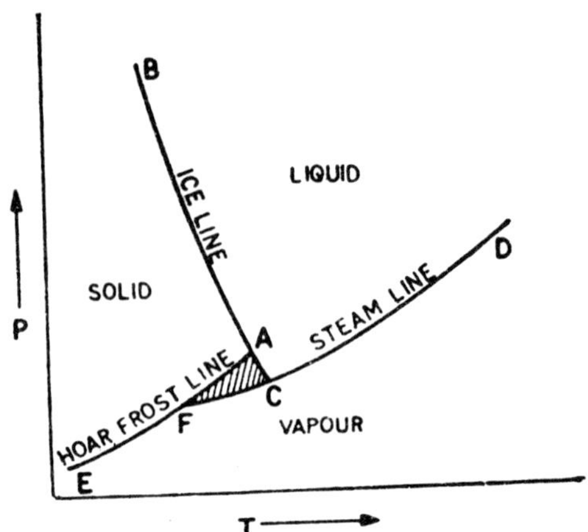

Fig. 4.16

state. Thus, in the shaded portion, the substance must exist simultaneously in the solid, liquid and the vapour states. It is not possible. Therefore, the three curves must meet at a single point O, called the triple point.

For water, the temperature and pressure corresponding to the triple point are 0.0075°C and 4.58 mm of Hg respectively. The triple point of water is not fixed but is different for different allotropic forms of ice.

Table 4.2 : Triple Point

	T(K)	Pressure (mm of Hg)
H_2	13.84	62.8
N_2	03.18	94.0
O_2	64.30	1.14
CO_2	216.65	3.880
H_2O	273.1676	4.58
I	387	90.0

4.26 GIBBS' PHASE RULE

Gibbs' Phase rule represents the condition of equilibrium between the number of coexisting phases and the components.

Consider a heterogeneous system consisting of P phases and containing C components in equilibrium. Let the different phases be represented by the superscripts a, b, c...p and let the components be represented by the subscripts 1, 2, 3,...C. The concentration of any of the components in the different phases may be different. The system is assumed to be in mechanical and thermal equilibrium. This means that all the phases are at the same temperature and pressure. It is further assumed that there is, no chemical reaction between the components in the system.

Let the chemical potentials of the different components in the P-phases be written as

$$\begin{array}{lll} \mu_1^a, \mu_2^a, \ldots & \ldots \mu C^a \\ \mu_1^b, \mu_2^b, \ldots & \ldots \mu C^b \\ \ldots & \ldots \ \ldots \\ \ldots & \ldots \ \ldots \\ \mu_1^p, \mu_2^p, \ldots & \ldots \mu C^p. \end{array}$$

Now, consider that small amounts (dn) of the components are transferred under equilibrium conditions, from one phase to another. It follows that

$$\left.\begin{array}{l} dn_1^a + dn_1^b + dn_1^c \ldots \quad \ldots dn_1^p = 0 \\ dn_2^a + dn_2^b + dn_2^c \ldots \quad \ldots dn_2^p = 0 \\ \ldots \quad \ldots \quad \ldots \quad \ldots \\ \ldots \quad \ldots \quad \ldots \quad \ldots \\ dn_C^a + dn_C^b + dn_C^c \ldots \quad \ldots dn_C^p = 0 \end{array}\right\} \quad \ldots(i)$$

From Gibbs' Duhem relation, at equilibrium, the change in Gibbs' potntial

$$dG = \Sigma\mu \ dn = 0$$

$$\left.\begin{array}{rl} \therefore \quad \Sigma\mu \ dn = & (\mu_1^a dn_1^a + \mu_1^b dn_1^b + \ldots \quad \ldots \mu_1^p dn_1^p) \\ & + (\mu_2^a dn_2^a + \mu_2^b dn_2^b + \ldots \quad \ldots \mu_2^p dn_2^p) \\ & \ldots \quad \ldots \quad \ldots \quad \ldots \\ & \ldots \quad \ldots \quad \ldots \quad \ldots \\ & + (\mu_C^a dn_C^a + \mu_C^b dn_C^b + \ldots \quad \mu_C^p dn_C^p) = 0 \end{array}\right\} \quad \ldots(ii)$$

Multiplying the set of equations in (i) by C different multipliers,

$\lambda_1, \lambda_2 \ldots \quad \ldots\lambda_c,$

$$\left.\begin{array}{llll}
\lambda_1 dn_1^a + \lambda_1 dn_1^b + & \ldots + \lambda_1 dn_1^p = 0 \\
\lambda_2 dn_2^a + \lambda_2 dn_2^b + & \ldots + \lambda_2 dn_2^p = 0 \\
\ldots \quad \ldots & \ldots \quad \ldots \\
\ldots \quad \ldots & \ldots \quad \ldots \\
\lambda_C dn_C^a + \lambda_C dn_C^b + & \ldots \lambda_C dn_C^p = 0
\end{array}\right\} \quad \ldots(iii)$$

Adding equations (ii) and (iii) and equating the coefficients of dn terms to zero, we get

$$\left.\begin{array}{l}
-\lambda_1 = \mu_1^b = \mu_1^b = \ldots \quad \ldots = \mu_1^p \\
-\lambda_2 = \mu_2^a = \mu_2^b = \ldots \quad \ldots = \mu_2^p \\
\ldots \quad \ldots \quad \ldots \\
\ldots \quad \ldots \quad \ldots \\
-\lambda_C = \mu_C^a = \mu_C^b = \ldots \quad \ldots = \mu_c^p
\end{array}\right\} \quad \ldots(iv)$$

The set of equations in (iv) are the equations of phase equilibrium. It means that at equilibrium, the chemical potentials of a component in all its phases must be equal. However, chemical potentials are different for different components.

In every phase, there are C different components having C different concentrations. But for a closed system if concentrations of (C – 1) components are arbitrarily chosen, the concentration of the last component is automatically fixed. Therefore, there are (C – 1) composition variables in each phase. For P phases the number of composition variables would be,

$$= P(C - 1).$$

In addition to this value, there are two other variables *viz.*, pressure and temperature. Thus, the total number of variables will be,

$$= P(C - 1) + 2. \quad \ldots(v)$$

Further, it has been shown that if chemical potential of any one of the P phases is defined then the chemical potential of the remaining (P – 1) phases is fixed. Therefore for C components the invariant composition variables will be

$$= C(P - 1). \quad \ldots(vi)$$

Therefore, the number of independent variables of a system will be given by

$$F = P(C - 1) + 2 - C(P - 1) \quad ...(vii)$$

Here F represents the number of degrees of freedom.

From equation (vii)

$$F = C - P + 2 \quad ...(viii)$$

or $$P + F = C + 2. \quad ...(ix\}$$

Water can exist in three states, *viz.*, solid, liquid and vapour. These are called the three phases of water. A component is defined as a distinct chemical substance different from others. Ice, water and steam refer to the same component viz., water.

In Fig. 4.15, above the curve CD, water exists only in the liquid state. In the space BOD, the pressure or temperature can be changed independently without any change in the phase of a substance. Thus a substance has two degrees of freedom (temperature and pressure) and only one phase (liquid).

Along the curve OD, the substance has two phases (liquid and vapour) but only one degree of freedom (temperature or pressure).

If the pressure is changed the temperature also changes along the curve OD.

At the triple point the substance has three phases (solid, liquid and vapour) and no degree of freedom.

According to Gibbs' phase rule, the sum of the number of phases and the number of degrees of freedom is equal to the number of components plus two.

Let P represent the number of phases, F the number of degrees of freedom and 0 the number of components. Then

$$P + F = C + 2$$

(i) For the liquid state, (region BOD]

$$P = 1, F = 2 \text{ and } C = 1$$

$$1 + 2 = 1 + 2$$

(ii) For the liquid-vapour state (curve OD)

$$P = 2, F = 1, C = 1$$

$$2 + 1 = 1 + 2$$

(iii) For the solid-liquid-vapour state (triple point)

$$P = 3, F == 0, C == 1$$

$$3 + 0 = 1 + 2.$$

The phase rule has practical applications when the components involved are more than one, *e.g.*, freezing mixtures (ice and salt).

4.27 HYGROMETRY

It is that branch of physics which deals with the study of water vapour in the atmosphere.

Relative Humidity

It is defined as the ratio of the mass of water vapour (m) actually present in a certain volume of air at room temperature to the mass of water vapour (M) required to saturate the same volume of air at the same temperature.

$$R.H. = \frac{m}{M}$$

Also $$R.H. = \frac{p}{P}$$ where P is the actual vapour pressure and P is the saturated vapour pressure at room temperature.

Dew Point

It is the temperature at which the water vapour actually present in the atmosphere is just sufficient to saturate it. It means that the actual vapour pressure at room temperature is equal to the saturated vapour pressure at dew point.

Absolute Humidity

It is the amount of water vapour actually present in one cubic metre volume of atmospheric air.

4.28 DENIELL'S HYGROMETER

Daniell's hygrometer consists of two glass bulbs A and S connected by a glass tubing and the space inside is evacuated and sealed (Fig. 4.17). The bulb A contains an evaporating liquid and the space above it, its vapour. The thermometer T_1 fixed on the stand measures the room temperature and the thermometer r, fixed inside the glass tubing dips in the liquid of bulb A and measures the dew point temperature.

A piece of cotton is tied around B and ether is poured over it. The vapours inside the bulb B condense, some vacuum space is created and the liquid evaporates and this causes cooling of the liquid in the bulb A. When the temperature of the bulb A reaches the dew point, the water vapours present in the atmospheric air condense on the bright metallic ring and it becomes dim. By continuously pouring ether on the bulb B, the temperatures at which dew just appears and disappears are noted and the mean temperature measures the dew point temperature.

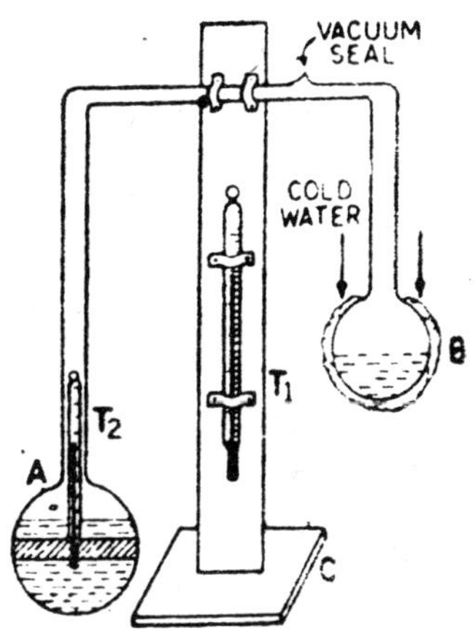

Fig. 4.17

From the standard tables the saturated vapour pressure at dew point and room temperature are noted and the relative humidity is calculated.

$$\text{RH} = \frac{\text{Saturated vapour pressure at dew point } (T_2)}{\text{Saturated vapour pressure at room temp.}(T_1)}$$

$$\%\ \text{RH} = \left(\frac{\text{Saturated vapour pressure at dew point } (T_2)}{\text{Saturated vapour pressure at room temp.}(T_1)}\right) \times 100$$

4.29 REGNAULT'S HYGROMETER

Regnault's hygrometer consists of two glass tubes A and B in which two thermometers T_1 and T_2 are fixed (Fig. 4.18). C_1 and C_2 are two bright silver caps fixed at the bottom of the glass tubes. One of the test tubes say A containing a volatile liquid (ether) is connected to an aspirator arid when water is drawn from the aspirator an equal amount of atmospheric air bubbles through the liquid in A and this causes evaporation of the liquid. The temperature of A goes down and when it reaches the dew point temperature dew appears on the metallic cap C_1 and the cap becomes dim. With the help of a telescope kept at a distance, the temperatures at which dew appears and disappears are noted and the mean of these two temperatures measures the dew point. The thermometer T_1 measures the room temperature.

From the standard tables, the saturated vapour pressures of water vapour corresponding to the room temperature and the dew point temperature are noted and relative humidity is calculated.

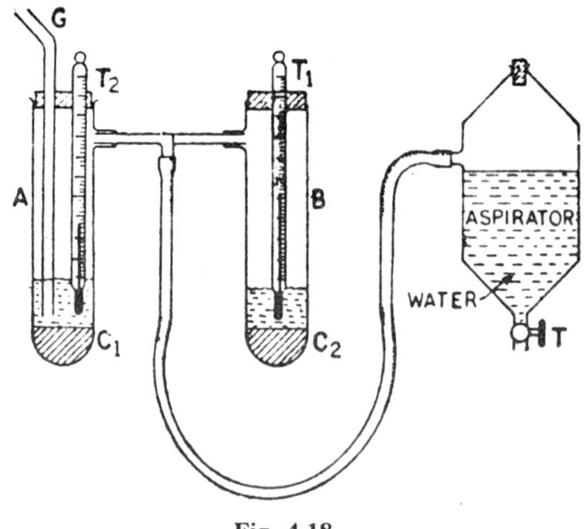

Fig. 4.18

$$\% \text{ RH} = \left[\frac{\text{Saturated vapour pressure at dew point } (T_2)}{\text{Saturated vapour pressure at room temp.} (T_1)}\right] \times 100$$

Advantages of Regnault's Hygrometer Over Daniell's Hygrometer

1. In the Regnault's hygrometer, the rate of cooling is controlled and is quicker as compared to the Daniell's Hygrometer.
2. In the Regnault's hygrometer, the air bubbling through the whole mass of ether keeps the same temperature throughout the liquid whereas in a Daniell's hygrometer the temperature at the surface of the liquid is less than the interior of the liquid.
3. Due to the silver caps at the bottom of the tubes in Regnault's hygrometer, temperatures inside and outside are the same. In the Daniell's hygrometer the metal ring is not in direct contact with the liquid due to the glass bulb.
4. In Regnault's hygrometer, ether is not wasted as in the case of Daniell's hygrometer.

5. In the case of Regnault's hygrometer, observations are taken at a distance with the help of telescope. Therefore, the breath of the observer and the heat of the body do not affect the humidity. It is not so m/the case of the Daniell's hygrometer.
6. Due to silver caps, the moment when the dew just appears or disappears can be noted very accurately in the case of the Regnault's hygrometer. But it is not so in the case of the Daniell's hygrometer.

4.30 WET AND DRY BULB HYGROMETER

Take two half degree centigrade thermometers. See that both how the same temperature when immersed in water. If they differ, change one of them. Take a piece of clean muslin cloth or wash the muslin cloth with soap. Remove any grease from the muslin piece. Finally, wash the piece with clean water. Tie one end of the muslin piece to the bulb of one of the thermometers. Suspend the thermometers from the iron rod of the retort stand (Fig. 4.19).

Adjust the height of the iron stand, so that one end of the muslin piece dips in water. The bulb of the thermometer should not dip in water. Water will rise through the muslin piece and will evaporate. Note the temperature of the dry bulb thermometer. Also record the temperature of the wet bulb thermometer (thermometer having muslin piece) after every one minute. Do not place the apparatus near an open window or under a fan. When the wet bulb thermometer shows a constant temperature, record it. Also note the temperature of the dry bulb thermometer after the experiment. Calculate RB and dew point with the help of constant tables.

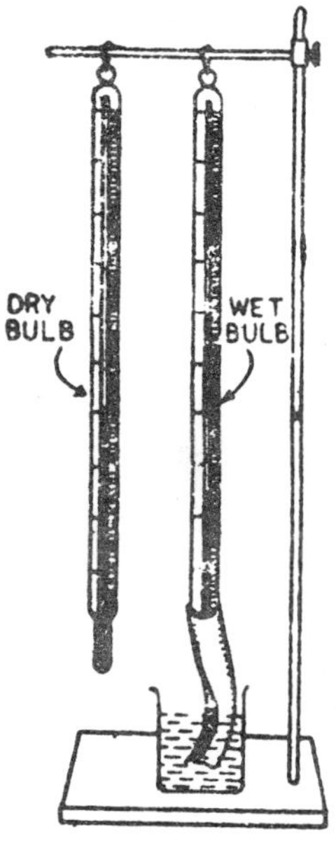

Fig. 4.19

Relative humidity is defined as the ratio of the actual vapour pressure at room temperature to the saturated vapour pressure at room temperature.

$$RH = \frac{\text{Actual VP at room temperature}}{\text{SVP at room temperature}}$$

Dew point is defined as that temperature at which water vapours actually present in the atmosphere are just sufficient to saturate it. The actual vapour pressure at room temperature is equal to the saturated vapour at dew point. Therefore,

$$\text{RH} = \frac{\text{Saturated VP at dew point}}{\text{Saturated VP at room temperature}}$$

Suppose, dry bulb reading before the experiment

$$= t_1 °C = \ldots$$

Dry bulb reading after the experiment

$$= t_2 °C = \ldots$$

$$\text{Mean dry bulb reading} = \left(\frac{t_{1+t_2}}{2}\right) °C.$$

No.	Wet bulb reading
1	...
2	...
...	
...	
...	
...	
...	
20	...

Calculations

Dry bulb reading A = ...

Wet bulb reading

(Constant reading at the end) B as ......

Difference between the dry and wet bulb readings

= (A – B) ....

From wet and dry bulb tumidity tables:

Saturated vapour pressure [Note the reading for dry bulb temperature (A) under zero column]

$$= P = ...$$

Actual vapour pressure [Note the reading for wet bulb temperature (B) under difference column for (A – B)]

$$= p = ...$$

$$\%RH = \frac{p}{P} \times 100.$$

Dew point (Find the value in the zero column for the value of p, and find the corresponding temperature)

$$= ..... \, ^{o}C$$

4.31 WEATHER

Dew, rain, clouds, fog, mist, hails, snow, hoar frost etc. are due to the condensation of water vapours present under different conditions in the atmosphere.

Dew

Dew is formed on grass and leaves in the early hours of the morning. The temperature of the atmosphere during the night gradually falls. In the early morning if the temperature of the atmosphere is less than the dew point temperature, formation of dew takes place. The saturated vapours when they come in contact with grass or leaves condense in the form of fine drops of water.

If the night is cloudy, the temperature of the earth does not fall and dew is not formed. A clear sky free from wind is helpful in the formation of dew.

Hoar Frost

If the temperature of the atmosphere falls below 0°C, the dew formed on the leaves and grass freezes and it is called hoar frost.

Fog and Mist

When water vapours present in the atmosphere condense on dust particles in the atmosphere, fog is formed. There .is no basic difference between fog and mist. Thick mist is called fog. A clear cold night is favourable for the formation of fog and mist.

Clouds and Rain

When the fog is formed at high altitudes it is called a cloud. When small drops of water in the cloud combine together, bigger water drops are formed and fall in the form of rain.

Snow and Hail

When the cloud suddenly passes through a very cold region, the water drops freeze and fall in the form of snow and hail. Snowfall is common in hill stations because of very low temperature in winter. Hails are caused due to the freezing of rain drops in the atmosphere. Hails are hard and of different sizes whereas snow is quite soft.

SOLVED EXAMPLES

Example 1.

If the temperature of air is 16.5°C and dew point is 6.5°C, find the % relative humidity of air.

S.V.P. at 6°, 7°, 16° and 17° are 7.05, 7.51, 13.62 and 14.42 mm respectively.

Solution:

Dew point = 6.5°C

Room temperature = 16.5°C

$$\text{S.V.P. at } 6.5°\text{C} = \frac{\text{S.V.P. at } 6° + \text{S.V.P. at } 7°}{2}$$

$$P = \frac{7.05 + 7.51}{2} = 7.28 \text{ mm}$$

$$\text{S.V.P. at } 165°\text{C} = \frac{\text{S.V.P. at } 16 + \text{S.V.P. at } 17}{2}$$

$$P = \frac{13.62 + 14.02}{2} = 14.2 \text{ mm}$$

$$\text{R.H.} = \frac{p}{P} \times 1000$$

$$= \frac{7.28}{14.02} \times 100 = 51.9\%$$

Example 2:

Calculate the dew point when the relative humidity of air at 20°C, is 52%. Given the S.V.P. of water at 20°C, 10°C and 9°C are 17.5, 9.2, 8.6 mm of Hg respectively.

Solution:

$$\text{R.H.} = \frac{\text{S.V.P. at dew point}}{\text{S.V.P. at room temperature}} = \frac{p}{P}$$

$$0.52 = \frac{p}{17.5}$$

$$p = 9.1 \text{ mm}$$

For 1°C difference = 9.2 – 8.6 = 0.6 mm

∴ Dew point temperature

$$= 9 + \frac{0.5}{0.6} = 9.83°\text{C}.$$

Example 3:

One kilogram of ice at –10°C is heated until the whale of it evaporates. How much heat is required? Latent heat of fusion of ice = 80 cal per gram and that of steam = 540 cal per gram.

Solution:

Mass of ice = 1 kg = 1000 g

Sp heat of ice = 0.5 cal/g-K

(1) Heat required to raise the temperature of ice from –10°C to 0°C = 1000 × 0.5 × 10 = 5000 cals

(2) Heat required to melt ice at 0°C

= 1000 × 80 = 80000 cals

(3) Heat required to raise the temperature of water from 0°C to 100°C = 1000 × 100 = 100000 cals

(4) Heat required to convert water into steam at 100°C

= 1000 × 540 = 540000 cals

Total quantity of heat required

$$= 5000 + 80000 + 10000 + 540000$$

$$= \mathbf{725000\ cals.}$$

Example 4:

One gram of steam is passed into a calorimeter containing 100 grams of ice. Calculate the amount of ice melted.

Solution:

Heat given out by one gram of steam when converted from steam into water at 100°C = mL = 1 × 537 = 537 cal.

Heat given out by one gram of water when cooled from 100°C to 0°C

$$= 1 \times 100 = 100 \text{ cal}$$

∴ Total quantity of heat given out by one gram of steam

$$= 537 + 100 = 637 \text{ cal}$$

Suppose the amount of ice melted

$$= m \text{ grams}$$

Latent heat of ice = 80 cal per gram

Heat absorbed by ice

$$= 80\ m \text{ cal}$$

Heat lost by steam = Heat gained by ice

$$637 = 80\ m$$

or $\mathbf{m = 7.96\ g}$

Example 5:

What will be the result of adding 52 g of ice to 100 g of water at, 40°C?

Solution:

Mass of water = 100 g

Temperature of water = 40°C

∴ Heat lost by water when its temperature falls from

$$40°C \text{ to } 0°C = 100 \times 40$$

$$= \mathbf{4000\ cal.}$$

If 52 g of ice melts, heat gained by it should be 52 × 80 = 4160 calories. But water car lose only 4000 calories of heat. Therefore, the whole of ice will not melt.

∴ Amount of heat gained by ice

$$= 4000 \text{ cal}$$

Amount of ice that melts

$$= \frac{4000}{80} = 50 \text{ g}$$

Remaining ice = 52 – 50 = 2 g

Hence, the result will be 2 g of ice and 150g of water at 0°C.

Example 6:

It takes fifteen minutes for an electric kettle to heat a certain quantity of water from 0°C to the boiling point 100°C. It requires 80 minutes to turn all the water at 100°C into steam. Deter- mine the latent heat of steam. (Assume the radiation losses to be negligible).

Solution:

Suppose mass of water = m

Heat required to raise its temperature from 0°C to 100°C

$$= 100 \text{ m}$$

∴ Heat produced by electric kettle in 15 minutes = 100 m.

Heat produced by electric kettle in 1 minute

$$= \frac{100\text{m}}{15}$$

Heat produced by electric kettle in 80 minutes

$$= \frac{100\text{m} \times 80}{15} \text{ cal}$$

Suppose latent heat of steam = L cal per gram

Heat required by m grams of water to boil off into steam

$$= \text{mL cal}$$

$$\therefore \qquad mL = \frac{100m \times 80}{15}$$

or $\qquad$ **L = 533.33 cal/g.**

Example 7:

A refrigerator converts 50g of water at 15°C in to ice at –20°C in one hour. Determine the quantity of heat removed per minute. (Specific heat of ice = 0.5)

Solution:

Here $\qquad m = 50$ g

Heat removed in cooling water from 15°C to 0°C

$= 50 \times 1 \times 15 = 750$ cal

Heat removed in converting water into ice at 0°C

$= 50 \times 80 = 4000$ cal

Heat removed in cooling ice from 0°C to –20°C

$= 50 \times 0.5 \times 20 = 500$ cal

Total heat removed in one hour

$= 750 + 4000 + 500 = 5250$ cal

Heat removed per minute

$$= \frac{5250}{60}$$

= 87.5 cal/minute.

Example 8:

20 g of a substance at 100°C, placed in the tube of a Bunsen's ice calorimeter, moves the mercury in the capillary tube 1 sq mm area through 50 mm. Find the specific heat of the substance if 1909 cc of water becomes 1099cc of ice on freezing.

Solution:

1000 cc of water becomes 1090 cc of ice on freezing.

It means 1000 g of ice on melting decreases in volume by

$$1090 - 1000 = 90\text{cc}$$

i.e., when one gram of ice melts decrease in volume

$$= \frac{90}{1000} = 0.09 \text{ cc}$$

Mass of the substance

$$= M = 20 \text{ g}$$

$$\theta = 100°C, \ S = ?$$

Area of cross-section = a = 1 sq mm

$$= \frac{1}{100} \text{ sq cm}$$

Length of the column through which liquid moves

$$= 50 \text{ mm} = 5 \text{ cm}$$

Decrease in volume = $v = l \times a = 5 \times 1/100$

$$= \frac{1}{20} \text{ cm}$$

$$MS\theta = \frac{80\,v}{0.09}$$

$$S = \frac{80\,v}{0.09 \times M \times \theta}$$

$$= \frac{80 \times 1}{20 \times 0.09 \times 20 \times 100}$$

$$= \mathbf{0.022 \ cal/g\text{-}0°C.}$$

Example 9:

A substance weighing 27 g and at 100°C was dropped into a Bunsen's ice calorimeter and the liquid column in the capillary tube recedes through a distance of 10 cm. Calculate the specific heat of the substance. Area of cross-section of the tube = 3sq mm.

Solution:

Here $M = 27$ g, $5 = ?$, $\theta = 100°C$

$l = 10$ cm,

$a = 3$ sq mm $= 3/100$ sq mm

$\therefore \qquad v = l \times a = 10 \times 3/100 = 3/10$ cc

$$MS\theta = \frac{80v}{0.09}$$

$$27 \times S \times 100 = \frac{80 \times 3}{10 \times 0.09}$$

$$\mathbf{S = 0.098\ cal/g^\circ C.}$$

EXERCISES

1. On a certain day the dew point is 8.5°C and the room temperature is 18.4°C. Find the KB if maximum vapour pressure for 8°C, 9°C, 18°C, and 19°C are 804, 8.61, 15.46 and 16.46 mm of mercury column respectively.
2. Find the result of mixing 10g of ice at –10°C with 10g of water at 10°C. Sp. heat of ice = 0.5 and latent heat of ice = 80cal/g.
3. What will be the result of mixing 400 g of copper chips at 500°C with 500g of melting ice?
5. The density of ice is 0-93 g/cc at 0.C. A piece of metal weighing 1.50 g is heated to 100°C and is then placed in a Bunsen's ice calorimeter. The decrease in volume is found to be 1.876 cc. Calculate the specific heat of the metal. Latent heat of ice = 80cal/g.
5. A solid of mass 20g is heated to a temperature of 80°C and quickly dropped in a Bunsen's ice calorimeter. The contraction in volume observed is 0.9 cc. The contraction in volume when one gram of ice melts is 0.09 cc. Latent heat of ice = 80cal/g. Calculate the specific heat of the solid.
6. Calculate how much steam from water boiling at 100°C will just melt 200g of wax at 15°C. (Melting point of wax = 55°C, Sq. heat of wax = 0.7, Latent heat of fusion of wax = 35cal per g).
7. What do you understand by change of state?
8. What is triple point 1 Show that the steam line, the ice line and the Hoar frost line meet at a single point.
9. Discuss Gibbs' phase rule. What do you understand by phase, degree of freedom and component?

10. Define hygrometry, relative humidity, dew point and absolute humidity.
11. Describe the construction and working of a Daniell's hygrometer.
12. Describe the construction and working of a Regnault's hygrometer. How is it used to determine the relative humidity? Compare it with a Daniell's hygrometer.
13. What is the difference between a gas and a vapour ?
14. Distinguish between saturated and unsaturated vapours.
15. What do you understand by relative humidity and dew point?
16. Describe briefly a wet and a dry bulb hygrometer.
17. Explain briefly the following:

    (i) Dew, (ii) Hoar frost, (iii) Fog, (iv) Mist, (v) Clouds, (vi) Snow, (vii) Hail, (viii) Rain.
18. Derive Gibbs' phase rule and discuss the equilibrium of a saturated solution and the solid of the dissolved substance.
19. On a certain day the RB is 66-67%. The saturated vapour pressure at room temperature is 18.6 mm. Calculate the saturated vapour pressure at dew point.
20. On a certain day the dew point is 18.6°C and the room temperature is 23.7°C. The saturated vapour pressures of water at 18°C, 19°C, 23°C, and 24°C are 15.46 mm, 16.46 mm, 21.02 mm and 22.32 mm of Hg respectively. Calculate the relative humidity.

5

Nature of Heat

5.1 INTRODUCTION

In the previous, chapters the various effects of heat and its properties have been discussed. At this stage it becomes necessary to formulate a cogent theory to explain the different heat phenomena and to see how far the proposed theory can explain the vast amount of data collected experimentally and otherwise. The theory should satisfactorily explain, primarily the following observed facts:

(a) thermal expansion,

(b) conservation of heat energy. When there is transfer of heat between two bodies the heat lost by one body is equal to the heat gained by the other,

(c) specific heat of bodies,

(d) thermal capacity of bodies,

(e) change of state from solid to liquid or liquid to gas,

(f) propagation of heat radiations in space with the velocity of light.

Two rival theories have been proposed to explain the properties and the effects of heat. The two theories are: *(i) Caloric theory of heat and (ii) Dynamic theory or the Kinetic theory of heat.* The caloric theory of heat assumes heat as some fluid and in the course of heat transfer from one body to another, this fluid is transferred. This theory held ground till the end of the eighteenth century.

The experiments of Joule, Rumford and Davy have shown that heat is a form of energy associated with the mechanical motion of material particles. Thus the kinetic theory of matter favoured the dynamical theory in preference to the caloric theory.

5.2 CALORIC THEORY OF HEAT

According to this theory heat is considered to be invisible, weightless fluid called *caloric*. This fluid can be produced when substances are burnt and this fluid can flow from a hot body to a cold body. The hot body is assumed to have more of this fluid and the cold body comparatively less. When a hot body is brought in contact with a cold body, the fluid flows from the hotter to the colder body, just as water flows from higher to lower level. This fluid is supposed to be all pervading, uncreatable and indestructible. The fluid is highly elastic and the particles of this fluid repel one another very strongly. When this fluid is added to a body its temperature rises and when the fluid is removed its temperature decreases. This fluid occupies the interstices of matter. Just as water can be obtained from a sponge by squeezing, heat can also be obtained from a body by extracting the caloric by friction. The caloric theory could explain in a crude and elementary way, different heat phenomena like conduction, thermal conductivity, latent heat etc.

5.3 FAILURE OF CALORIC THEORY

The caloric theory held ground till the end of the eighteenth century. The work of Rumford, an artillery engineer began to cast doubts about the validity of the caloric theory. Rumford was engaged in the supervision of boring of cannon for the government of Bavaria. The boring process produced tremendous amount of heat and to prevent this overheating the cannon was immersed in water. Even this water boiled off during the process and the boring process had to be stopped. According to the caloric theory, it was admitted that the caloric had to be supplied to water to raise its temperature to the boiling point. The continuous extraction of caloric during the process of boring was explained on the hypothesis that when matter was finely subdivided its capacity to hold the caloric decreases, and the caloric thus released causes the water to boil.

It was interesting to note that the water surrounding the cannon continued to boil even after the borer became blunt and when there was no further cutting. This proves that even a blunt drill continuously produced an inexhaustible supply of heat which is contrary to the explanation given on the basis of the caloric theory. Further, while explaining the production of heat the conservation principle which is very vital to the concepts of physics has not been kept in mind. During the boring process, mechanical work was continuously expended and caloric was continuously extracted. Rumford's observation gave him an opportunity of eliminating the two cases of non-conservation. This process of elimination is possible only when heat also is considered as a form of energy. What is involved is not the disappearance of the mechanical energy and the appearance of heat energy but it is the transformation of one form of energy into another. Rumford made some measurements regarding the work done and the corresponding heat produced, though his experiments lacked precision. It was later the work of Joule during the years 1843 to 1878, that established the equivalence of heat and work as two forms of energy.

5.4 MECHANICAL EQUIVALENT OF HEAT (JOULE'S EXPERIMENTS)

James Prescott Joule (1818-1889) performed a series of experiments to prove convincingly that heat is also a form of energy. His findings that there is a definite equivalence between the work done and the corresponding amount of heat and any other form of energy, put an end to the controversy that existed between the caloric theory and the dynamical theory of heat. The dynamical theory of heat was firmly established.

He developed the argument that if an amount of work W (or any other form of energy) disappears a definite quantity of heat H is produced.

$$W \propto H$$

or

$$W = JH$$

$$J = W/H.$$

The constant J is called the mechanical equivalent of heat. The mechanical equivalent of heat (J) is defined as the amount of work done to produce a unit quantity (1 calorie) of heat. The value of J in C.G.S. system is $4.18\ 0 \times 10^7$ ergs/caloric. In the M.K.S. system, its value is 4.18 joules/calorie.

Joule's method for determining the mechanical equivalent of heat: The apparatus consists of a calorimeter with the fixed blades a, a etc. fixed to the calorimeter radially. S is a spindle which is fitted with the vanes b, b radially and these vanes are called the movable vanes (Fig. 5.1).

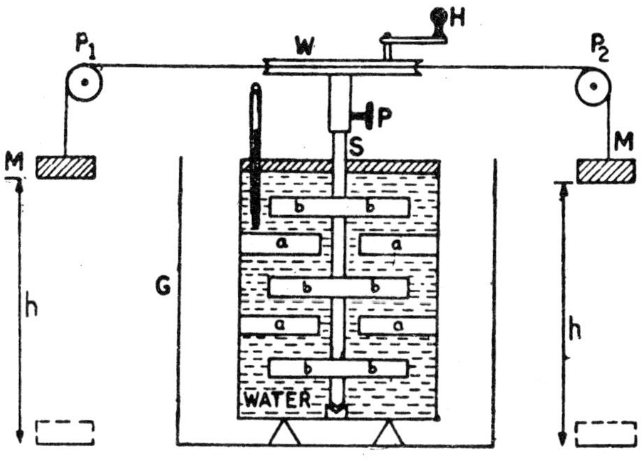

Fig. 5.1

When the spindle rotates, the, vanes a, a are stationary and the vanes b, b rotate. The water between any two vanes gets warmed up due to the frictional forces between the two vanes.

W is a, wheel with a handle B, which can rotate about a vertical axis. Two strings are wound in the same direction over the drum W and these strings pass over the pulleys P_1 and P_2 and they carry the masses M, M at the other ends.

When the masses are moved up and left free, they move down due to gravity, thus rotating the drum and the spindle. When the masses are moved up the pin P is removed and when the masses move down the pin P is fixed. This is done because when the masses are moved up, the work done is not taken into calculation.

Let the masses move down by a distance of h cm

Loss of potential energy of the two masses a = 2 Mgh

If the masses fall down n times, the loss of energy

$= n \times 2Mgh.$

Let the velocity of the masses at the end of the fall be v cm/s

To find v, the time taken by the masses to fall through a height h is noted. The average velocity = h/t Therefore the velocity at the end of the fall will be twice the average velocity because the initial velocity is zero.

$$\therefore \qquad v = 2\left(\frac{h}{t}\right)$$

Kinetic energy of the two masses

$$= 2 \times 1/2\ Mv^2$$

$$= Mv^2.$$

For n times the KE $= n \times Mv^2$

$\therefore$ Energy used up in heating the water and the calorimeter,

$$W = n[2Mgh - Mv^2]$$

Let the total water equivalent of the calorimeter, water, fixed vanes, movable vanes etc. be w

Initial temperature $= \theta_1$

Final temperature $= \theta_2$

Heat produced $= H$

$$= w(\theta_2 - \theta_1)$$

$\therefore$ J the mechanical equivalent of heat

$$= W/H.$$

$$= \frac{n[2Mgh - Mv^2]}{w(\theta_2 - \theta_1)}$$

The value of J as a mean of a number of observations is taken. The value of J is 4.18×10^7 ergs/cal.

5.5 ROWLAND'S EXPERIMENT

Rowland found the value of J accurately in the year 1879. In the Joule's original experiment there were two main defects, (i) The rate of rise of temperature was small [about 0.6°C per hour] and so the loss of heat due to radiation was appreciable and (ii) the observed temperatures were not corrected to the standard gas scale.

Rowland's apparatus consists of a calorimeter C fitted with vanes on its inner side (Fig. 5.2). The calorimeter is suspended from a torsion wire. D is a drum over which cords are wound. The ends of the cords are connected to the masses M and M, passing over the pulleys. The cords leave the drum at diametrically opposite points. Thus a couple is applied on the calorimeter. AB is a rod having sliding masses m and m and are used to increase the moment of inertia of the system.

The spindle S has paddles attached to it and it is rotated with a very high speed with the help of a steam engine. The counters C_2 and C_2 are used to measure the number of rotations of the spindle. A sensitive thermometer T already corrected with a standard gas scale is used to measure the temperature of water in the calorimeter, The calorimeter is surrounded by a water jacket in which Water is circulated at constant temperature to find the cooling correction.

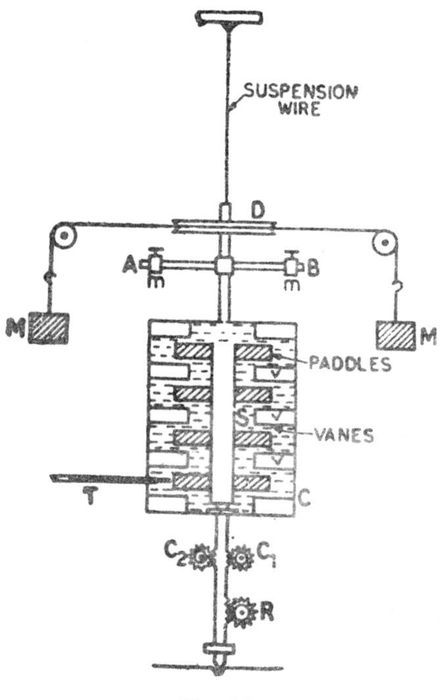

Fig. 5.2

The initial temperature of water in the calorimeter is noted. The spindle is set rotating and the masses M, M are adjusted quickly so that the calorimeter does not rotate and remains in the equilibrium position. The rotation is continued till there is sufficient rise in temperature. Suppose the initial temperature of water is θ_1 and after n rotations of the spindle, ifs temperature is θ_2. The rotations are measured with the help of the counters. The couple applied by the masses = MgD. Here D a the diameter of the drum.

For n rotations, the work done,

$$W = 2\ \pi n.MgD$$

Heat produced, $H = (m + w)\ (\theta_2 - \theta_1)$.

Here m is the mass of water and w is the water equivalent of the calorimeter, the vanes, the paddles and the spindle.

$$\therefore \qquad J = \frac{W}{H}$$

$$= \frac{2\pi n\, MgD}{(m + w)\ (\theta_2 - \theta_1)} \text{ ergs/cal}.$$

Following corrections are to be applied (i) cooling correction and (ii) correction for the torsional couple, *i.e.*, the weight Mg in air, should be reduced to vacuum.

The results obtained by Rowland at various temperatures were as follows:

Temperature °C	**J (joules per calorie)**
10	4.196
15	4.188
20	4.181
25	4.176
30	4.174
35	4.175

The mean value of J is taken as 4.18 joules per caloric
or $\qquad 4.18 \times 10^7$ ergs per caloric.

5.6 SEARLE'S FRICTION CONE METHOD

The apparatus consists of two truncated friction cones A and S made of brass. The cone B is fixed to an ebonite disc C fitted inside a brass cylinder E. F is an ebonite ring which holds B in position. S is a spindle attached to the base of the cylinder S. The spindle is rotated at high speed by hand or with an electric motor. The number of rotations made by the spindle is measured with the help of the counters C_1 and C_2 (Fig. 5.3).

A grooved wooden disc D is fixed to the inner cone A with the help of pins passing through the iron-ring I. The iron ring also provides the necessary weight for the cone A to be in its position. A cord passes around the groove of the disc D and the other end is connected to a mass M. The chord passes over a pulley A mercury thermometer calibrated to gas scale is used to measure the temperature of water in the cone A.

The outer cone B is rotated with the help of the spindle and it rends to rotate the cone A. The mass M is adjusted quickly so that the cone A does not rotate. A couple Mgr is provided by the mass in the opposite direction. Due to friction, heat is produced and water in the cone A gets heated.

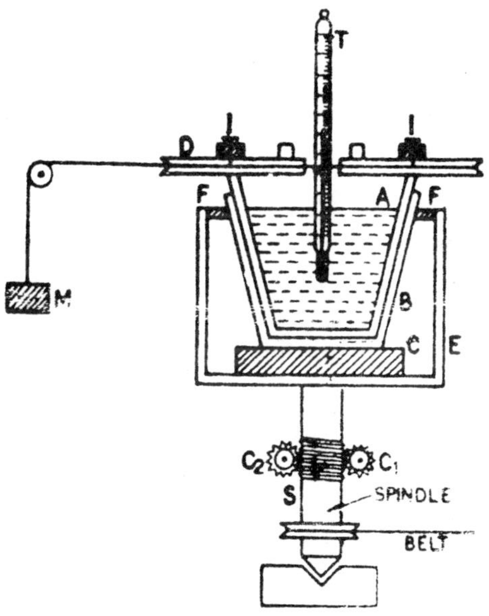

Fig. 5.3

Suppose,

Initial temperature of water $= \theta_1$

Final temperature of water $= \theta_2$

Rise in temperature $= (\theta_2 - \theta_1)$

Number of rotations made by the spindle $= n$

Mass of water $= m$

Water equivalent $= w$

Radius of the disc D $= r$

Work done, $W = 2\pi n\,(Mgr)$

Heat produced $H = (m + w)\,(\theta_2 - \theta_1)$

$$\therefore \quad J = \frac{W}{H}$$

$$= \frac{2\pi n\, Mgr}{(m + w)\,(\theta_2 - \theta_1)} \qquad \text{...(i)}$$

Radiation Correction: The time for which the spindle in rotated is noted. The fall in temperature is noted for the same time after the experiment. Let $\delta\theta$ be the fall in temperature. Then radiation correction $= (\delta\theta/2)$

$\therefore$ The correct rise in temperature

$$= \left[(\theta_2 - \theta_1) + \frac{\delta\theta}{2}\right]$$

$$\therefore \quad J = \frac{2\pi n\ Mgr}{(m + w)\left[(\theta_2 - \theta_1) + \dfrac{\delta\theta}{2}\right]} \text{ ergs/cal}$$

5.7 JOULE'S CALORIMETER (DETERMINATION OF J)

The apparatus consists of a copper calorimeter containing water to about 2/3 of its volume. A resistance coil R is Immersed in water and the two ends of R are connected to two binding terminals fixed to the lid. A stirrer and a thermometer are inserted through holes in the lid.

The calorimeter is closed in a wooden box to minimize the loss of heat (Fig. 5.4).

The electrical connections are shown in Fig. 5.4. If a current of I amperes is passed through the resistance coil R at a potential difference of V volts for a time t seconds, then the amount of energy liberated $= VIt + 10^7$ ergs.

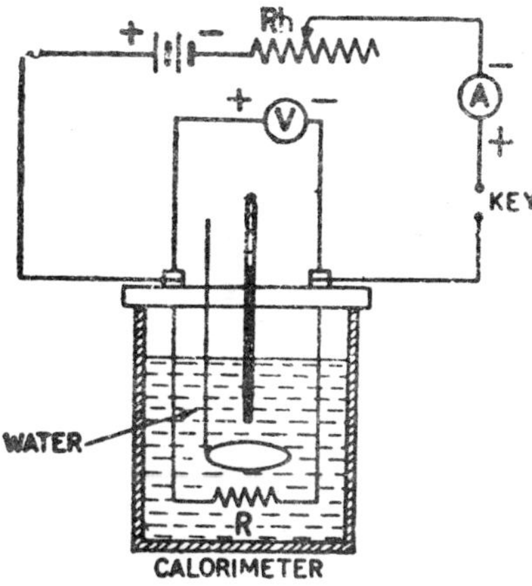

Fig. 5.4

Knowing the total water equivalent w, and the initial and final temperatures of the calorimeter θ_1 and θ_2, the heat produced B can be calculated

$$H = w(\theta_2 - \theta_1).$$

To apply the radiation correction, the calorimeter is allowed to cool for the same interval of time t as that of the passage of current,, and the fall in temperature θ_3 is noted. Half of this fall in temperature is added to the final observed temperature θ_2.

$\therefore$ Corrected final temperature $= \theta_2 + \dfrac{\theta_3}{2}$

Mechanical equivalent of heat

$$J = \frac{\text{Work done}}{\text{Heat produced}}$$

$$J = \frac{VIt \times 10^7}{w\left(\theta_2 + \dfrac{\theta_3}{2} - \theta_1\right)} \text{ ergs/cal}$$

In case, the resistance of the voltmeter is not high, the actual value of current passing through R is calculated, knowing the values of R and the reading of the voltmeter. The drawback of this experiment is the small rate of rise of temperature due to low currents that can be drawn from accumulators used as the source of supply.

5.8 CALLENDAR AND BARNES CONT'NUOUS FLOW METHOD (DETERMINATION OF J)

The Callendar and Barnes apparatus consists of a resistance coil R enclosed inside a narrow glass tube (Fig. 5.5). The two ends of the wire are connected to two thick copper connectors. Continuous flow of water can be maintained through the tube and the temperatures of inlet and outlet Water are measured with thermometers T_1 and T_2. The central tube containing R is surrounded by a vacuum jacket which minimises the loss of heat by conduction and convection. The vacuum jacket is surrounded by an outer jacket through which water is circulated continuously. This ensures the loss of heat by radiation to be steady.

The coil R is connected in series with a battery, a rheostat, an ammeter and a key. A voltmeter is connected across the terminals of R. For more accurate work, the current through the coil and the potential difference across its ends are measured using a potentiometer, initially calibrated with a standard cell.

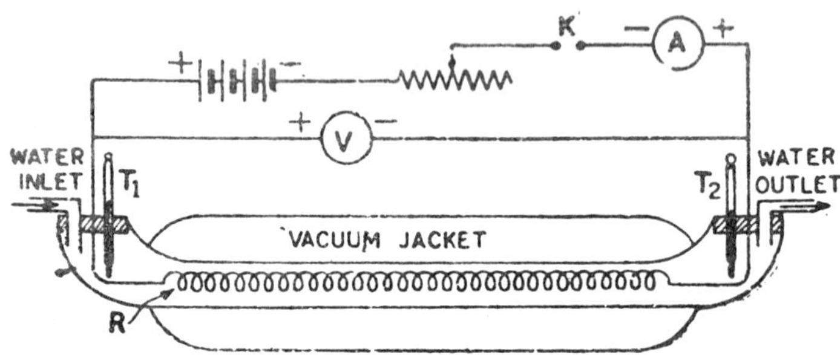

Fig. 5.5 : Callendar and Barnes' Continuous Flow Apparatus.

When the rate of now of water and the current piling through R are maintained constant, a steady state will be reached and the thermometers T_1 and T_2 will show constant readings. At this stage, the temperatures of every parts of the apparatus will remain steady. When the thermometers T_1 and T_2 show constant readings, say θ_1 and θ_2, the mass of water (m) flowing in a time t seconds is measured. If S is the mean value of the specific heat of water between the temperatures θ_1 and θ_2 and H the quantity of heat lost by radiation etc., then the amount of heat produced by passage of electric current through the coil.

$$= mS\,(\theta_2 - \theta_1) + H. \quad \quad ...(i)$$

If V and I are the voltmeter and ammeter readings then the amount of energy liberated in time t second. = VIt × 10^7 ergs. The corresponding quantity of heat produced

$$= \frac{VIt \times 10^7}{J} \text{ cal} \qquad \text{...(ii)}$$

From (i) and (ii)

$$= \frac{VIt \times 10^7}{J} = mS(\theta_2 - \theta_1) + H. \qquad \text{...(iii)}$$

To eliminate H, a second set of observations is taken by altering the values of the current and the rate of flow of water such that the difference of temperature $\theta_2 - \theta_1$ remains the same after the steady state is reached. If V', I' and m' are the corresponding values for the same time t, then

$$\frac{V'I't \times 10^7}{J} = m'S(\theta_2 - \theta_1) + H. \qquad \text{...(iv)}$$

As the temperature difference $(\theta_2 - \theta_1)$ is maintained the same in the two cases, the temperatures of the various parts of the apparatus will be the same and hence the loss of heat by radiation etc. will also be the same.

Subtracting (iv) from (iii)

$$\frac{(VI - V'I)t \times 10^7}{J} = (m - m')(\theta_2 - \theta_1)$$

$$J = \frac{(VI - V'I')t \times 10^7}{(m - m')S(\theta_2 - \theta_1)} \text{ ergs/cal}$$

All the quantities in the above expression can be measured very accurately. With calibrated platinum resistance thermometers, the temperature difference can be measured with great precision. Similarly, the values of the current and the potential difference can also be measured very accurately. Thus, this method ensures an accurate determination of the value of J.

5.9 JAEGAR AND STEINWEHR'S METHOD

In 1921 Jaegar and Steinwehr performed an experiment to find the value of j accurately. Their method has an accuracy of 1 in 1,000. The value of J obtained by them is equal to 4.186 × 10^7 ergs per calorie.

The apparatus consists of a large cylinder containing about 50 kg of water. R is a heater coil and T is a sensitive platinum resistance thermometer calibrated to the gas scale. S_1 and S_2 are two electrically rotating stirrers (Fig. 5.6). The cylinder is surrounded by an outer jacket through which water is circulated at constant temperature. This will help in estimating the cooling correction accurately. The current is measured with a standard resistance placed in series with the heating resistance R. The potential difference across the standard resistance is obtained with a sensitive potentiometer initially calibrated with a standard cell. The time for which the current is passed is measured accurately with a chronograph.

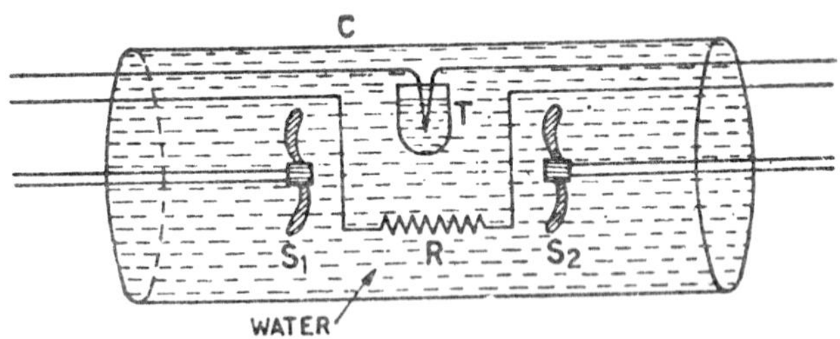

Fig. 5.6

Let the potential difference across R be V volts and the current flowing for t seconds be I amperes.

Work done $= VIt$ joules

Heat produced $= (m + w)(\theta_2 + \theta_1)$

Here m is the mass of water and w is the water equivalent of the apparatus.

$$J = \frac{VIt}{(m + w)(\theta_2 - \theta_1)} \text{ joules/calorie}$$

or

$$J = \frac{VI \times 10^7}{(m - W)\, S\, (\theta_2 - \theta_1)} \text{ ergs/calorie}$$

This method has the following advantages:

1. The mass of water used is very large. Hence the correction for the thermal capacity of the vessel is small.

2. A small rise in temperature of the order of 1.4°C is obtained. This will enable to apply the cooling correction accurately because Newton's law of cooling is applicable.
3. Efficient stirring of water is provided. Thus the temperature of water and the calorimeter remains uniform.
4. As the surrounding temperature of the cylinder is kept constant by circulating water at constant temperature, cooling correction can be accurately determined.

5.10 KINETIC THEORY OF MATTER

The experiments on the conversion of work into heat have shown clearly that heat is a form of energy. This energy is connected with the motion of molecules of which the matter is made of Kinetic theory of matter gives an explanation to the nature of this motion and the nature of heat energy.

According to Kinetic theory of matter, every substance (in the form of solid, liquid or gas) consists of a very large number of very small particles called the molecules. The molecules are the smallest particles of a substance that can exist in free state. The molecules possess the characteristic properties of the parent substance. The molecules are in a state of continuous motion with all possible velocities. The velocity of the molecules increases with rise in temperature.

The energy possessed by the molecules can be of two forms, kinetic or potential. When the stop-cock of an evacuated flask is opened, air rushes in quickly to fill in the space. This shows that the molecules possess rapid motion and hence they possess kinetic energy. When a solid expands on heating, the molecules are pulled apart against the forces of intermolecular expansion. The amount of work done in separating the molecules to larger distances manifests itself as the potential energy of the molecules. Thus, the amount of heat given to a solid substance increases the energy of the molecules and this increase in energy is partly kinetic and partly potential.

Thus, the kinetic theory of matter is based on the following three points:

(i) matter is made up of molecules,

(ii) molecules are in rapid motion, and

(iii) molecules experience forces of attraction between one another.

The validity and the soundness of any theory depends on its capability to explain the observed facts, viz., the three states of matter, expansion, change of state and conduction.

5.11 THE THREE STATES OF MATTER

Solid

In the case of a solid the intermolecular distances are small and hence the forces of intermolecular attraction are large. The only motion permissible to a molecule is oscillation or vibration about its mean position which is fixed. Hence a solid has define size and shape. Heating a solid means giving more energy to the molecules. This will enable the molecules to vibrate more violently about their mean position. The molecules move further apart and this results in the expansion of the solid. With the supply of more heat to the solid, the amplitude of vibration of the molecules in- creases to such an extent that the molecules will be free to leave their mean position because the kinetic energy of the molecules is- greater than the potential energy due to the forces of intermolecular attraction. Now the molecules are free to move about within the body of the substance and this corresponds to the liquid state of the substance. The amount of heat supplied, which is equivalent to the work done in pulling the molecules apart against the forces of intermolecular attraction, determines the latent heat of fusion of the solid.

Liquid

In a liquid, the molecules are farther apart than the solid. The molecules are not confined to any fixed mean position. They are free to move about within the volume of the liquid. Hence the liquids have their own size but no shape.

The definite size of the liquid is due to the mutual forces of attraction between the molecules. The molecules present on the surface of the liquid experience a resultant downward force due to the presence of the molecules present within the sphere of influence and hence the surface of a liquid behaves as a stretched membrane. This is the phenomenon of surface tension. The volume of the liquid is fixed at a given temperature but the molecular distance are comparatively larger than that of a solid. Therefore the molecules experience less force of attraction and they can occupy any position within the liquid. Hence the liquid occupies the shape of the containing vessel.

The molecules are always in a state of random motion. They collide with each other and they may or may not change their respective kinetic energies. As already explained, every molecule on the surface of the liquid is held in its position as a result of the forces of intermolecular attraction and surface tension. Due to the collisions a certain molecule may acquire more energy than the energy with which it is held. As soon as this happens, the molecule falls off, *i.e.*, it leaves the liquid surface. This is called evaporation. If no external heat is supplied it is obvious that the mean kinetic energy of the molecules decreases and the temperature of the liquid falls. This explains cooling caused by evaporation. If the moving vapour molecules come near to each other at sufficiently low velocities, they experience a greater magnitude of force of attraction. The molecules coalesce and this results in condensation. Rate of evaporation increases with increase in temperature of the liquid and decrease in pressure over the liquid surface.

Gas

In the case of a gas, the intermolecular distances are much larger than that of a solid or a liquid and the molecules of a gas are free to move about in the entire space available to them. Hence a gas has no shape or size. If the space occupied by a gas is increased, the gas will occupy the whole of the new space uniformly. This is due to the fact that the molecules possess rapid random motion and the force of intermolecular attraction is negligibly small. The molecules move about independently in straight lines and the only restriction to this movement is collision with another molecule or collision with the walls of the containing vessel. In a gas the more predominant factor is the movement of the molecules than the force of attraction between the molecules.

5.12 CONCEPT OF IDEAL OR PERFECT GAS

At very low pressures, the forces of intermolecular attraction are negligible. Therefore, at low pressures only, a real gas obeys the equation PV = RT. An ideal gas can be defined as a real gas at low pressure. For a real gas the internal energy is a function of pressure and temperature. By definition, the ideal gas should satisfy the equation

$$PV = RT \qquad ...(i)$$

and $$\left[\frac{dU}{dP}\right]_T = 0 \qquad ...(ii)$$

$$\left(\frac{dU}{dV}\right)_T = \left[\frac{dU}{dP}\right]_T \left[\frac{dP}{dV}\right]_T$$

From equation (i)

$$\frac{dP}{dV} = -\frac{RT}{V^2}$$

$$= -\left[\frac{RT}{V}\right]\left[\frac{1}{V}\right] = -\frac{P}{V}$$

$$\therefore \quad \left(\frac{dU}{dV}\right)_T = \left[\frac{dU}{dP}\right]_T \left[-\frac{P}{V}\right]_T \qquad ...(iii)$$

As $\left(-\frac{P}{V}\right)$ is not equal to zero and

$$\left[\frac{dU}{dP}\right]_T = 0_S$$

$$\therefore \quad \left[\frac{dU}{dV}\right]_T = 0 \qquad ...(iv)$$

Since for an ideal gas $\left[\frac{dU}{dP}\right]_T$ and $\left[\frac{dU}{dV}\right]_T$ are both equal to zero, the internal energy of an ideal gas is a function of temperature only.

$$\therefore \quad U = f(T) \qquad ...(v)$$

In a large number of calculations, a real gas can be taken approximately as an ideal gas. This involves an error that can be tolerated in the calculations. For all practical purposes, real gases below a pressure of two atmospheres can be considered as ideal gases. A saturated vapour in equilibrium with its own liquid can be considered to have the properties of an ideal gas. The error involved is low.

5.13 KINETIC THEORY OF GASES

The continuous collision of the molecules of the gas with the walls of the containing vessel and their reflection from the walls results in the change of momentum of the molecules. According to Newton's Second Law of motion, the rate of change of momentum per unit area of the wall surface corresponds to the force exerted by the gas per unit area. The force per unit area measures the pressure of the gas.

Postulates of the Kinetic Theory of Gases

(i) The gas is composed of small indivisible particles called molecules. The properties of the individual molecules are the same as that of the gas as a whole.

(ii) The distance between the molecules is large as compared to that of a solid or liquid and hence the forces of intermolecular attraction are negligible.

(iii) The molecules are continuously in motion with varying velocities and the molecules move in straight lines between any two consecutive collisions. The collisions do not alter the molecular density of the gas, *i.e.*, on the average the number of molecules present in a unit volume remains the same. Also, the molecules do not accumulate at any place within the volume of the gas.

(iv) The size of the molecules is infinitesimally small as compared to the average distance traversed by a molecule between any two consecutive collisions. The distance between any two consecutive collisions is called free path and the average distance is called the mean free path. The mean free path is dependent on the pressure of the gas. If the pressure is high the mean free path is less and if the pressure is low the mean free path is more.

(v) The time of impact is negligible in comparison to the time taken to traverse the free path.

(vi) The molecules are perfectly hard elastic spheres and the whole of their energy is kinetic.

5.14 EXPRESSION FOR THE PRESSURE OF A GAS

As mentioned earlier, the continuous impact of the molecules on the walls of the containing vessel accounts for the pressure of the gas.

Consider a cubical vessel ABCDEFGH of side l cm containing the gas (Fig. 5.7). The volume of the vessel and hence that of the gas is l^3cc. Let n and m represent the very large number of molecules present in the vessel and the mass of each molecule respectively.

Consider a molecule P moving in a random direction with a velocity C_1. The velocity can be resolved into three perpendicular components u_1, v_1 and w_1 along the X, Y and Z axes respectively.

Therefore, $C_1^2 = u_1^2 + v_1^2 + w_1^2$

The component of the velocity with which the molecule P will strike the opposite face BCFG is u_1 and the momentum of the molecule is mu_1. This molecule is reflected back with the same momentum mv_1 in an opposite direction and after traversing a distance l will strike the opposite face ADEH.

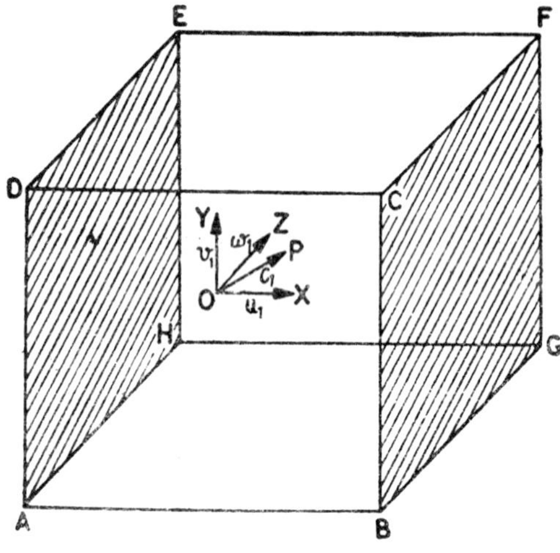

Fig. 5.7

The change in momentum produced due to the impact is

$$mu_1 - (-mu_1) = 2mu_1.$$

As the velocity of the molecule is u_1, the time interval between two successive impacts on the wall BCFG is

$$\frac{2l}{u_1} \text{ seconds}$$

$\therefore$ No. of impacts per second

$$= \frac{1}{\frac{2l}{u_1}}$$

$$= \frac{u_1}{2l}$$

Change in momentum produced in one second due to the impact of this molecule is

$$2mu_1 \times \frac{u_1}{2l} = \frac{mu_1^2}{l}$$

The force F_x due to the impact of all the n molecules in one second

$$= \frac{m}{l}\left[u_1^2 + u_2^2 + ... + u_n^2\right]$$

Force per unit area on the wall BCFG or ADEH is equal to the pressure P_X

$$P_X = \frac{m}{l \times l^2}(u_1^2 + u_2^2 + u_3^2 + ... + u_n^2)$$

Similarly the pressure P_Y on the walls CDEF and ABGH is given by

$$P_Y = \frac{m}{l^3}(v_1^2 + v_2^2 + ... + v_n^2)$$

and the pressure P_Z on the walls ABCD and EFGH is given by

$$P_Z = \frac{m}{l^3}(w_1^2 + w_2^2 + ... + w_n^2)$$

As the pressure of a gas is the same in all directions, the mean pressure P is given by

$$P = \frac{P_X + P_Y + P_Z}{3}$$

$$= \frac{m}{3l^3}\Big[(u_1^2 + v_1^2 + w_1^2) + (u_2^2 + v_2^2 + w_2^2)$$

$$+ (u_3^2 + v_3^2 + w_3^2) + ... + (u_n^2 + v_n^2 + w_n^2)\Big]$$

$$= \frac{m}{3l^3}\left[C_1^2 + C_2^2 + C_3^2 + ... C_n^2\right] \qquad ...(i)$$

But volume, $V = l^3$. Let C be the root-mean-square velocity of the molecules (R.M.S. velocity).

Then $\quad C^2 = \dfrac{C_1^2 + C_2^2 + C_3^2 + ... C_n^2}{n}$

or $\quad nC^2 = C_1^2 + C_2^2 + C_3^2 + ...C_n^2$

Substituting this value in equation (i), we get

$$P \frac{m.nC^2}{3V} \qquad ...(ii)$$

But M = mn where M is the mass of the gas of volume V, m is the mass of each molecule and n is the number of molecules in a volume V.

$$\therefore \qquad P \frac{MC^2}{3V}$$

$$\text{or} \qquad P \frac{1}{3}\rho C^2 \qquad ...(iii)$$

$$\because \qquad \frac{M}{V} = \rho \text{ the density of the gas.}$$

From equation (iii)

$$C^2 = \frac{3P}{\rho}$$

$$\text{or} \qquad C = \sqrt{\frac{3P}{\rho}}$$

[**Note:** R.M.S. velocity C is the square root of the mean of the squares of the individual velocities and it is not equal to the mean velocity of the molecules.]

Table : Molecular Velocities at 0°C

Gas	Molecular Weight	Root mean square velocity in cm/s
Hydrogen	2.016	18.4×10^4
Helium	4	13.1×10^4
Nitrogen	28	4.95×10^4
Oxygen	32	4.01×10^4
Argon	40	4.14×10^4
Carbon dioxide	44	3.95×10^4
Chlorine	71	3.11×10^4

R.M.S. Velocity of Hydrogen

The density of hydrogen at N.T.P. is 0.000089 g/cc. Therefore, C for hydrogen can be calculated as follows:

$$C = \sqrt{\frac{3P}{\rho}}$$

$$= \sqrt{\frac{3 \times 76 \times 13.6 \times 981}{0.000089}}$$

$$= 1.84 \times 10^5 \text{ cm/s.}$$

(b) For oxygen the density at N.T.P.

$$= 16 \times 0.000089$$

$$\therefore \text{ C for oxygen} = \sqrt{\frac{3 \times 76 \times 13.6 \times 981}{16 \times 0.000089}}$$

$$= 4.6 \times 10^4 \text{ cm/s.}$$

(c) For air the density at N.T.P. is 0.001293 g/cc

$$\therefore \qquad \text{C for air} = \sqrt{\frac{3 \times 76 \times 13.6 \times 981}{0.001293}}$$

$$= 4.850 \times 10^4 \text{ cm/s.}$$

3.15 KINETIC ENERGY PER UNIT VOLUME OF A GAS

$$P = \frac{1}{3} \rho C^2$$

$$= \frac{2}{3} \cdot \frac{1}{2} \rho C^2$$

$$= \frac{2}{3} . E.$$

where, $E = 1/2 \ \rho C^2$ and is equal to the kinetic energy per unit volume of the gas ρ is the mass per unit volume. Hence, the pressure of a gas is numerically equal to two-thirds of the mean kinetic energy of translation of a unit volume of the molecules.

5.16 KINETIC INTERPRETATION OF TEMPERATURE

The pressure of a gas, according to the kinetic theory, is

$$P = \frac{1}{3} \rho C^2$$

$$P = \frac{1}{3}\frac{MC^2}{V}$$

$$PV = \frac{1}{3}MC^2$$

Consider 1 gram molecule of the gas at a temperature T K

$$PV = RT$$

$$\therefore \quad 1/3\ MC^2 = RT$$

$$\frac{1}{2}MC^2 = \frac{3}{2}RT \qquad ...(i)$$

Let the mass of each molecule be m and Avogadro's number be N.

$$M = m \times N$$

$$\frac{1}{2}mNC^2 = \frac{3}{2}RT$$

$$\frac{1}{2}mC^2 = \frac{3}{2}\frac{R}{N}T$$

$$= \frac{3}{2}kT \qquad ...(ii)$$

Here if is called Boltzmann's Constant.

Thus, from equation (ii), the mean kinetic energy of a molecule is directly proportional to the absolute temperature of a gas. When the temperature of the gas is increased, the mean kinetic energy of the molecules increases. When heat is withdrawn from a gas, the mean kinetic energy of the molecules decreases. Thus, at absolute zero temperature, the kinetic energy should be zero. It means at absolute zero temperature, the molecules are in a perfect state of rest and have no kinetic energy. But before the absolute zero temperature is reached, all gases change their state to liquids and solids.

Also from equation (ii), $C^2 \propto T$

It means that the root mean square velocity of the molecules is also directly proportional to the square root of the absolute temperature.

5.17 DERIVATION OF GAS EQUATION

From kinetic theory

$$P = \frac{1}{3}\rho C^2$$

$$P = \frac{1}{3}\frac{M}{V}C^2$$

$$PV = \frac{1}{3}MC^2$$

Consider one gram molecule of a gas at an absolute temperature T. The mean energy of the molecules

$$= \frac{1}{2}MC^2$$

$$= \frac{1}{2}NmC^2$$

$$\therefore \quad PV = \frac{1}{3}NmC^2$$

$$= \frac{2}{3}N.\frac{1}{2}.mC^2$$

Mean kinetic energy of a molecule

$$= \frac{1}{2}mC^2$$

$$= \frac{3}{2}kT$$

$$\therefore \quad PV = \frac{2}{3}N.\frac{3}{2}kT$$

$$PV = NkT$$

But $N \times k = R$

$$\therefore \quad PV = RT \qquad \text{...(ii)}$$

Note on the gas equation.

In the gas equation

$PV = RT$

P is in dynes/sq cm

$R = 8.31 \times 10^7$ ergs/g mol-K

T is in K

V is the volume in cc per gram molecule.

5.18 DERIVATION OF GAS LAWS

(i) Boyle's Law

According to the kinetic theory,

$$P = \frac{1}{3}\rho C^2$$

$$P = \frac{1}{3}\frac{M}{V}C^2$$

$$PV = \frac{1}{3}C^2. \qquad ...(i)$$

At a constant temperature T, C^2 is constant. Therefore at a constant temperature

$$1/3\ MC^2 = \text{constant}$$

Hence $\quad PV = \text{const.}$ –at constant temperature.

(ii) Charles' Law

According tot he kinetic theory

$$P = \frac{1}{3}\rho C^2$$

$$P = \frac{1}{3}\frac{M}{V}C^2$$

$$PV = \frac{1}{3}MC^2.$$

Consider one gram molecule of a gas at absolute temperature T.

$$M = mN$$

$$PV = \frac{1}{3}NmC^2. \qquad ...(i)$$

The mean kinetic energy of a molecule

$$\frac{1}{2}mC^2 = \frac{3}{2}kT$$

$$mC^2 = 3kT$$

Substituting this value in equation (i),

$$\therefore \qquad PV = NkT \qquad ...(ii)$$

where N is the Avogadro's number and k is the Boltzmann's constant.

If $\qquad$ P is constant,

$$V \propto T$$

It means that for a given mass of gas, the volume is directly proportional to the absolute temperature provided the pressure remains constant. This is Charles' Law.

(iii) **Regnault's Law**

From equation (ii)

$$PV = NkT$$

When V is constant,

$$P \propto T.$$

It means that for a given mass of gas, the pressure is directly proportional to its absolute temperature provided the volume remains constant.

5.19 AVOGADRO'S HYPOTHESIS

Consider two gases A and B at a pressure P and each having a volume V.

Mass of each molecule of the first gas = m_1

Number of molecules of the first gas = n_1

Mean square velocity of the molecules of the first gas = C_1^2

For the first gas

$$P = \frac{1}{3} \rho_1 C_1^2$$

$$= \frac{1}{3} \frac{m_1 n_1 C_1^2}{V} \quad \text{...(i)}$$

Similarly for the second gas

$$P = \frac{1}{3} \rho_2 C_2^2$$

$$= \frac{1}{3} \frac{m_2 n_2 C_2^2}{V} \quad \text{...(ii)}$$

where, m_2 represents the mass of each molecule, n the number of molecules and C_2^2 the mean square velocity of the molecules.

From (i) and (ii),

$$\frac{1}{3} = \frac{m_1 n_1 C_1^2}{V} = \frac{1}{3}\frac{m_2 n_2 C_2^2}{V} \qquad \text{...(iii)}$$

or $$m_1 n_1 C_1^2 = m_2 n_2 C_2^2$$

If the two gases are at the same temperature T, the mean kinetic energy of the molecules of both the gases is the same

$$\frac{1}{2} m_1 C_1^2 = \frac{1}{2} m_2 C_2^2$$

$$\therefore \qquad m_1 C_1^2 = m_2 C_2^2 \qquad \text{...(iv)}$$

From (iii) and (ii),

$$n_1 = n_2$$

Hence, equal volumes of all gases under similar conditions of temperature and pressure have the same number of molecules. This represents Avogadro's hypothesis.

5.20 GRAHAM'S LAW OF DIFFUSION OF GASES

According to the kinetic theory

$$P = \frac{1}{3}\rho C^2$$

or $$C = \sqrt{\frac{3P}{\rho}}$$

$$C \propto \frac{1}{\sqrt{\rho}}.$$

It means that the root mean square velocity of the molecules of a gas is inversely proportional to the square root of its density. Consider two gases whose root mean square velocities are C_1 and C_2.

$$\frac{C_1}{C_2} = \sqrt{\frac{\rho_2}{\rho_1}}$$

or $$\frac{r_1}{r_2} = \sqrt{\frac{\rho_2}{\rho_1}}$$

Here r_1 and r_2 represent the rates of diffusion of the two gases.

5.21 DEGREES OF FREEDOM AND MAXWELL'S LAW OF EQUIPARTITION OF ENERGY

A molecule in a gas can move along any of the three co-ordinate axes. It has three degrees of freedom. Degrees of freedom mean the number of independent variables that must be known to describe the state or the position of the body completely. A monoatomic gas molecule has three degrees of freedom. A diatomic gas molecule has three degrees of freedom of translation and two degree of freedom of rotation. It has in all five degrees of freedom.

According to kinetic theory of gases, the mean kinetic energy of a molecule at a temperature T is given by

$$\frac{1}{2}mC^2 = \frac{3}{2}kT \qquad ...(i)$$

But $$C^2 = u^2 + v^2 + w^2$$

As x, y and z are all equivalent, mean square velocities along the three axes are equal

$\therefore$ $$u^2 + v^2 = w^2$$

or $$\frac{1}{2}m(u^2) = \frac{1}{2}m(v^2) = \frac{1}{2}m(w^2)$$

$\therefore$ $$\frac{1}{2}mC^2 = 3\left[\frac{1}{2}m(u^2)\right] 3\left[\frac{1}{2}m(v^2)\right]$$

$$= 3\left[\frac{1}{2}m(w^2)\right]$$

$$= \frac{3}{2}kT$$

$\therefore$ $$\frac{1}{2}mu^2 = \frac{1}{2}kT \qquad ...(ii)$$

$$\frac{1}{2}mv^2 = \frac{1}{2}kT \qquad ...(iii)$$

$$\frac{1}{2}mw^2 = \frac{1}{2}kT \qquad ...(iv)$$

Therefore, the average kinetic energy associated with each degree of freedom = 1/2 kT.

Thus, the energy associated with each degree of freedom (whether translatory or rotatory) is 1/2 kT.

This represents the theorem of equipartition of energy.

5.22 ATOMICITY OF GASES

(i) Monoatomic Gas: A mono-atomic gas molecule has one atom. Each molecule has three degrees of freedom due to translatory motion only.

Energy associated with each degree of freedom = 1/2 kT

Energy associated with three degrees of freedom = 3/2 kT

Consider one gram molecule of a gas.

Energy associated with one gram molecule of a gas

$$= N \times 3/2\ kT$$

$$= 3/2\ (N \times k)\ T$$

[But, $N \times k = R$

$$\therefore \qquad U = \frac{3}{2} RT.$$

This energy of the gas is due to the energy of its molecules. It is called internal energy U. For an ideal gas, it depends upon temperature only.

$$\therefore \qquad C_v = \frac{dU}{dT} = \frac{3}{2} R$$

(dU/dT is the internal energy per unit degree rise of temperature]

But $$C_P - C_V = R$$

$$C_P = C_V + R$$

$$= \frac{3}{2} R + R = \frac{5}{2} R$$

For a mono-atomic gas

$$\gamma = \frac{C_P}{C_V}$$

$$= \frac{\frac{5}{2} R}{\frac{3}{2} R} = 1.67$$

The value of γ is found to be true experimentally for monoatomic gases like argon and helium.

(ii) *Diatomic Gas:* A diatomic gas molecule has two atoms. Such a molecule has three degrees of freedom of translation and two degrees of freedom of rotation.

Energy associated with each degree of freedom

$$= \frac{1}{2}kT$$

Energy associated with 5 degrees of freedom $= \frac{5}{2}kT$

Consider one gram molecule of gas.

Energy associated with 1 gram molecule of a diatomic gas.

$$= N \times \frac{5}{2}\mathbf{kT} = \frac{5}{2}RT$$

$$U = \frac{5}{2}RT$$

$$C_V = \frac{dU}{dT}$$

$$= \frac{5}{2}R$$

But $\quad C_P - C_V = R$

$$C_P = C_V + R$$

$$= \frac{5}{2}R + R = \frac{7}{2}R$$

$$\gamma = \frac{C_P}{C_V}$$

$$= \frac{7/2\ R}{5/2\ R} = 1.40.$$

The value of $\gamma = 1.40$ has been found to be true experimentally for diatomic gases like hydrogen, oxygen, nitrogen etc.

(iii) *Triatomic Gas*: (a) A triatomic gas having 6 degrees of freedom has an energy associated with 1 gram molecule

$$= N \times \frac{6}{2}kT = 3RT$$

$$U = 3RT$$

$$C_V = \frac{dU}{dT} = 3R$$

But $$C_P - C_V = R$$

$$C_P = C_V + R$$

$$= 3R + R = 4R$$

$$\gamma = \frac{C_P}{C_V}$$

$$= \frac{4R}{3R} = 1.33.$$

(b) A triatomic gas having 7 degrees of freedom has an energy associated with one gram molecule

$$= N \times \frac{7}{2} kT = \frac{7}{2} R$$

$$U = 7/2\ RT$$

$$C_V = \frac{dU}{dT} = \frac{7}{2} R$$

But $$C_P - C_V = R$$

$$C_P = C_V + R$$

$$= \frac{7}{2} R + R = \frac{9}{2} R$$

$$\gamma = \frac{C_P}{C_V}$$

$$= \frac{9/2\ R}{7/2\ R} = 1.28.$$

Thus, the value of γ, C_p and C_v can be calculated depending upon the degrees of freedom of a gas molecule.

5.23 MAXWELL'S LAW OF DISTRIBUTION OF VELOCITY

At a particular temperature, a gas molecule has a fixed mean kinetic energy. It does not mean that the molecule is moving with the same speed throughout its movement. After each encounter the speed of the molecule changes and due to a large number of collisions, the speed is different at different instants. But the root mean square velocity (r.m.s.) C remains the same at a fixed temperature. At any instant, all the molecules are not moving with the same velocity. Some are moving with a velocity higher than C and the others with a velocity lower than C. But the mean kinetic energy of all the molecules remains constant at a given temperature.

Derivation of Maxwell's Law of Distribution of Molecular Velocities

The mean square velocity of molecules is defined by the equation

$$C^2 = \frac{1}{N}\int_0^\infty c^2 \, dN .$$

Here dN is the number of molecules having velocities between c and c + dc. If the total number of molecules is N, then a fraction $\frac{dN}{N}$ will have the components of velocities in x direction in the range u and u + du. However, the fraction $\frac{dN_x}{N}$ is a function of U only and is proportional to du.

$$\therefore \qquad \frac{dN_x}{N} = f(u)\, du$$

$$\therefore \qquad dN_x = N\, f(u)\, du$$

The equation for y and z directions are

$$dN_y = N\, f(v)\, dv$$

and $$dN_z = N\, f(w)\, dw.$$

Also the value of N is too large and du is small compared with u but in range u and u + du, there are a large number of molecules. The fraction of dN_x molecules whose y components of velocity lie in the range v and v + dv is given by the equation

$$\frac{d^2N_{x,y}}{dN_x} = \frac{dN_y}{N} = f(v)\, dv$$

$$\therefore \qquad d^2N_{z,y} = N\, f(u)\, f(v) du\, dv$$

The number of molecules represented by $d^2N_{x,y}$ is still large. Suppose that the fraction of these molecules $d^2N_{x,y,z}$ whose components of velocity in Z direction in the range w and w + dw is given by the equation

$$d^3N_{x,y',z} = Nf(u)\ f(v)\ f(w)du\ dv\ dw$$

The density of velocity points is given by the equation

$$\rho = \frac{d^3N_{x,z,y}}{du\ dv\ dw} = Nf(u)\ f(v)\ f(w)$$

As there is no preferred direction of the velocities, the density of velocity points ρ is constant *i.e.*, velocity space is isotropic.

Also $\qquad C^2 = u^2 + v^2 + w^2$

$\therefore \qquad Nf(u)\ f(v)\ f(w) = \text{constant} \qquad$ ...(i)

when $\qquad u^2 + v^2 + w^2 = \text{constant}. \qquad$ ...(iii)

These equations (i) and (ii) must be simultaneously satisfied. Equation (ii) limits the values of the variables in equation (i) and reduces the number of independent variables to two. Therefore, it is called an equation of constraint.

Differentiating equations (i) and (ii)

$$f(v)\ f(w)\frac{\partial f(u)}{\partial u}du + fd(w)\ f(u)\frac{\partial f(v)}{\partial v}dv + f(u)\ f(v)\frac{\partial f(w)}{\partial w}dw = 0 \qquad \text{...(iii)}$$

and $\qquad 2u\ du + 2v\ dv + 2w\ dw = 0 \qquad$ ...(iv)

Simplifying equations (iii) and (iv)

$$\frac{1}{f(u)}\frac{\partial f(u)}{\partial u}du + \frac{1}{f(v)}\frac{\partial f(v)}{\partial v}dv + \frac{1}{f(w)}\frac{\partial f(w)}{\partial w}dw = 0 \qquad \text{...(v)}$$

and $\qquad udu + udv + udw = 0. \qquad$ ...(vi)

As there are only two independent variables, arbitrary values cannot be given to all the three differentials. Suppose the arbitrary values are given to dv and dw. Multiply equation (vi) by mβ where m is the mass of the molecules and β is an arbitrary, unknown function. The product mβ is called Langrange undetermined multiplier.

$$m\beta\ [udu + vdv + wdw] = 0 \qquad \text{...(vii)}$$

Adding equations (v) and (vii)

$$\left[\frac{1}{f(u)}\frac{\partial f(u)}{\partial u} + m\beta u\right] du + \left[\frac{1}{f(v)}\frac{\partial f(v)}{\partial v} + m\beta v\right] dv$$

$$+\left[\frac{1}{f(w)}\frac{\partial f(w)}{\partial w} + m\beta w\right] dw\ 0 \qquad \text{...(viii)}$$

As the two variables v and w are considered to be independent, their differentials are arbitrary and the values, dv = 0 and dw = 0 may be assigned to them.

From equation (viii)

$$\left(\frac{1}{f(u)}\frac{\partial f(u)}{\partial u} + m\beta u\right) = 0 \qquad \text{...(ix)}$$

Taking dv = 0, and dw ≠ 0, we get

$$\frac{1}{f(w)}\frac{\partial f(w)}{\partial w} + m\beta w = 0 \qquad \text{...(x)}$$

Taking dw = 0 and dv ≠ 0

$$\frac{1}{f(v)}\frac{\partial f(v)}{\partial v} + m\beta v = 0 \qquad \text{...(xi)}$$

To satisfy equations (ix), (x) and (xi), β would either be a constant or a function of the variable in that equation. A constant value of β is the only value that can satisfy all the equations. It is clear from the above equations that the function f satisfies the same differential equation irrespective of the component.

For the x component

$$\frac{d}{du}[\log f(u)] = -m\beta u$$

$$\int d[\log f(u)] = -m\ \beta \int u\ du$$

$$\log f(u) = -\beta\ (1/2\ mu^2) + \log A \qquad \text{...(xii)}$$

Here A is a constant of integration.

From equation (xii)

$$f(u) = Ae^{-\beta(1/2\ mu^2)} \qquad \text{...(xiii)}$$

Similarly $f(v) = Ae^{-\beta(1/2\ mv^2)}$ ...(xiv)

and $f(w) = Ae^{-\beta(1/2\ mw^2)}$ ...(xv)

From these three equations, the density of velocity points is given by

$$\rho = N A^3 e^{\beta}\left[\frac{1}{2} m (u^2 + v^2 + w^2)\right]$$

or $$\rho = N A^3 e^{-\beta}\left[\frac{1}{2} m c^2\right] \quad ...(xvi)$$

But $$\rho = \frac{d^3N_{x,y,z}}{du\, dv\, dw}$$

i.e., the number of molecules with components of velocity lying in a small cubical volume divided by the volume element. However ρ is constant within the infinitesimal spherical shell within the radii c and c + dc. The volume of this clement is $4\pi c^2 dc$. This volume element contains dN molecules.

$$\therefore \quad \rho = \frac{dN}{4\pi c^2\, dc} \quad ...(xviii)$$

From equations (xvi) and (xviii)

$$\frac{dN}{4\pi c^2 dc} = NA^3\, e^{-\beta\left[1/2\, mc^2\right]}$$

or $$\frac{dN}{dc} = N\pi NA^3 c^2 e^{-\beta\left[1/2\, mc^2\right]} \quad ...(xix)$$

$$\therefore \quad dN = 4\pi NA^3 e^{-bc^2} c^2 dc \quad ...(xx)$$

where N represents the number of molecules per cc and

$$A = \sqrt{\frac{b}{\pi}} \quad \text{and} \quad b = \frac{\beta m}{2}$$

and $$b = \frac{m}{2kT},\ \beta = \frac{1}{kT}$$

Here k is Boltzmann's constant.

The graph representing dN/dc and speed is shown in Fig. 5.8.

The shaded area in the figure represents the number of molecules dN having velocities between c and c + dc. The number of molecules possessing the root mean square velocity C is maximum.

The number of molecules having very low or very high velocity is small. The total area under the curve represents the total number of molecules.

From equation (xx)

$$\frac{dN}{dc} = 4\pi N \left[\frac{m}{2\pi kT}\right]^{3/2} c^2 e^{-\frac{mc^2}{2kT}}$$

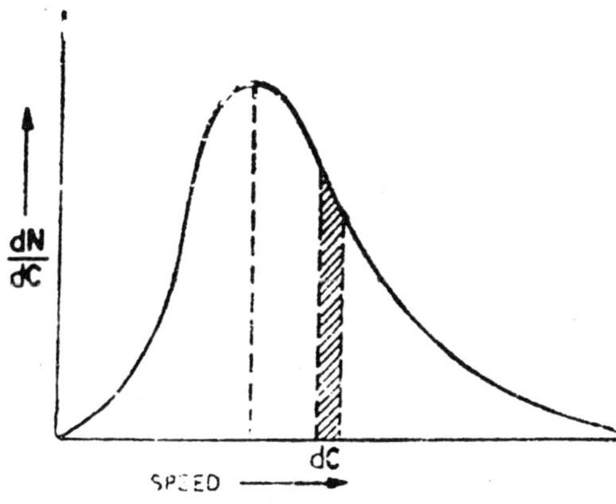

Fig. 5.8

Taking 1/2 mc^2 = E, where E is the mean kinetic energy of a molecule,

$$\frac{dN}{dc} = \frac{4N}{\sqrt{\pi}}\left(\frac{m}{2kT}\right)^{3/2} c^2 e^{-\frac{mc^2}{2kT}} \qquad \text{...(xxi)}$$

Taking 1/2 mc^2 = E where E is the mean kinetic energy of a molecule,

$$\frac{dN}{dc} = \frac{4N}{\sqrt{\pi}}\left(\frac{m}{2kT}\right)^{3/2} c^2 e^{-\frac{E}{kT}} \qquad \text{...(xxii)}$$

Thus, Maxwell's law of distribution of velocities shows that the mean kinetic energy of all the molecules of a gas remains constant at a fixed temperature, though at any instant the molecules are moving with different velocities.

5.24 EXPERIMENTAL VERIFICATION OF VELOCITY DISTRIBUTION

Zartman and Ko (1934) performed an experiment to study the distribution of velocity. The apparatus consists of an oven with an opening at A. S_1 and S_2 are two parallel slits. A drum D can be rotated about an axis passing through O. P is a glass surface on which a beam of silver atoms will get deposited (Fig. 5.9).

Metallic silver is melted in the oven. A beam of silver atoms is ejected through A. The drum D is rotated with a speed of 6000 r.p.m. approximately. When the drum is stationary, the silver atoms get deposited at the same point on the glass plate. When the drum rotated a fine beam of silver molecules enters through the slit S_2. Molecules with very high speeds reach the plate P first, *i.e.*, on the right end Sf P and molecules with very low speeds reach the other end of the plate. After a short time, sufficient quantity of silver is deposited on the plate P. Using a spectro-photometer, the relative intensity of silver on the plate P is studied and this represents the velocity distribution of the molecules. The graph representing the number of molecules and velocity agrees with Maxwell's distribution of velocity.

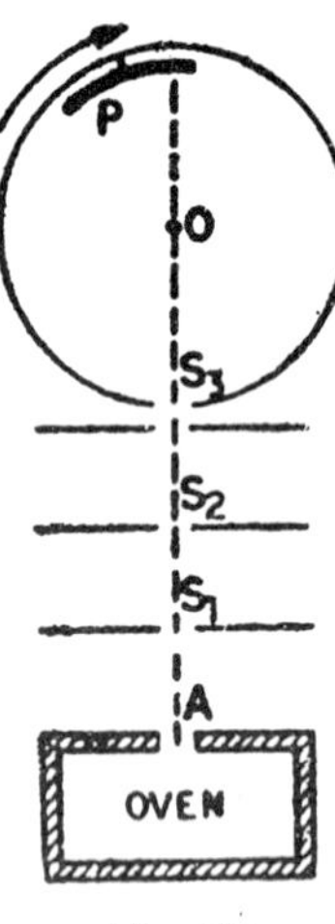

Fig. 5.9

In 1947, Estermann, Simpson and Stern designed a more precise apparatus to study the velocity distribution.

Cesium atoms from the oven emerge from the opening A (Fig. 5.10). S is a slit and D is a hot tungsten wire. The whole apparatus is enclosed in an evacuated chamber (pressure 10^{-8} mm of Hg). The opening A and the slit S are horizontal. In the absence of a gravitational field, cesium

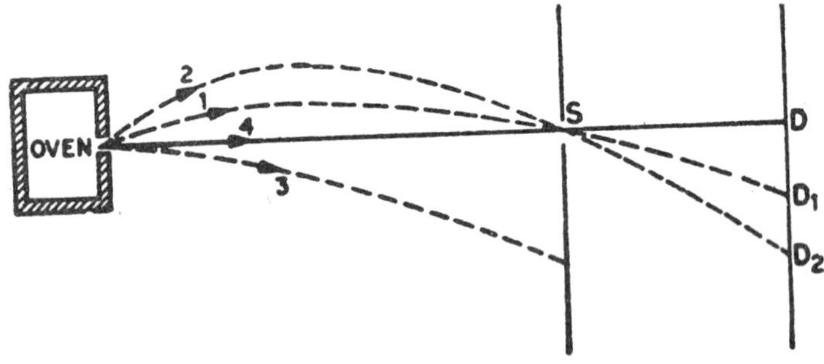

Fig. 5.10

atoms will strike the wire at D. But due to the gravitational field, the path is a parabola. The atoms going along the path 3 do not reach the wire. The atoms going along the paths 1 and 2 reach at D_1 and D_2 respectively. The velocity of the atoms m path 1 is higher than the path 2.

When cesium atoms strike the wire they get ionized and re-evaporate. They are collected by a-negatively charged detecting cylinder surrounding the tungsten wire. The magnitude of the current indicates the intensity of-the atoms at various positions. The detector can be moved to different positions of the wire. The atoms reaching at D_1 have higher velocity than those reaching at D_2. The vertical height of the detector represents the magnitude of the velocity and the ionization current indicates the number of atoms striking the wire at a particular point. A graph is drawn between the ionization current along the y-axis and the vertical height (speed of the atoms) of the detector along the x-axis. The velocity distribution is found to be in agreement with the Maxwellian distribution law of velocity.

5.25 MEAN FREE PATH

In deriving the expression for the pressure of a gas on the basis of kinetic theory, it was assumed that the molecules are of negligible size. They were assumed to be geometrical points. A geometrical point has

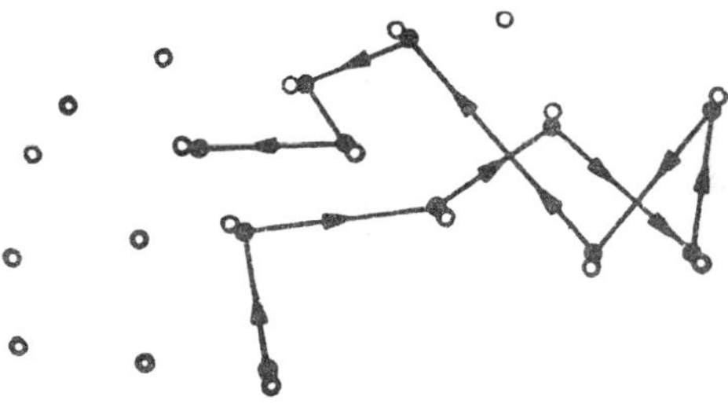

Fig. 5.11

no dimensions and hence intermolecular collisions will not be possible. But, a molecule has a finite size (though small) and moves in the space of the vessel containing it. It collides with other molecules and the walls of the containing vessel The path covered by a molecule between any two consecutive collisions is a straight line, and is called the *free path*. The direction of the molecule is changed after every collision. After a number of collisions, the total path appears to be zig-zag and the free path is not constant (Fig. 5.11). Therefore, a term *mean free path* is used to indicate the mean distance travelled by a molecule between two collisions. If the total distance travelled after N collisions is S, then the mean free path λ is given by

$$\lambda - \frac{S}{N} \qquad ...(i)$$

Let the molecules be assumed to be spheres of diameter d. A collision between two molecules will take place if the distance between the centres of the two molecules is d. Collision will also occur if the colliding molecule has a diameter 2d and the other molecule is simply a geometrical point. Thus, assuming all other molecules to be geometrical points and the colliding molecule of diameter 2d, this molecule will cover a volume $\pi d^2 v$ in one second. This corresponds to the volume of a cylinder of diameter 2d and length v.

Let n be the number of molecules per cc.

Then, the number of molecules present in a volume $\pi d^2 v$

$$= \pi d^2 v \times n$$

This value also represents the number of collisions made by the molecule in one second.

The distance moved in one second = v and the number of collisions in one second = = $\pi d^2 v \times n$.

$$\therefore \text{ Mean free path } \lambda = \frac{v}{\pi d^2 vn}$$

$$= \frac{1}{\pi d^2 n} \qquad ...(i)$$

This equation was deduced by Clausius,

$$\therefore \qquad \lambda \propto \frac{1}{d^2} \qquad ...(ii)$$

The mean free path is inversely proportional to the square of the diameter of the molecules.

Let m be the mass of each molecule.

Then, $\qquad m \times n = \rho$

$$\lambda = \frac{m}{\pi d^2 \rho} \qquad ...(iii)$$

The mean free path is inversely proportional to the density of the gas.

The expression for the mean path according to Boltzmann is

$$\lambda = \frac{3}{4\pi d^2 n} \quad ...(iv)$$

He assumed that all molecules have the same average speed. Maxwell derived the expression,

$$\lambda = \frac{1}{\sqrt{2}\,.\,\pi d^2 n} \quad ...(v)$$

He calculated the value of A on the basis of the law of distribution of velocities

Mean Free Path (λ)

Gas	d	λ
Hydrogen	2.47×10^{-3} cm	1.83×10^{-5} cm
Nitrogen	3.50×10^{-3} cm	0.944×10^{-5} cm
Helium	2.18×10^{-3} cm	2.85×10^{-5} cm
Oxyzen	3.39×10^{-3} cm	0.999×10^{-5} cm

Determination of Mean Free Path

The relation between coefficient of viscosity and the mean free path of a molecule is given by

$$\eta = \frac{1}{3} mnC\lambda$$

For unit volume,

$$mn = \rho$$

$$\therefore \quad \eta = \frac{1}{3} \rho C\lambda$$

$$\lambda = \frac{3\eta}{\rho C} \quad ...(i)$$

The root mean square velocity C of a molecule can be calculated knowing pressure, density and temperature. The coefficient of viscosity of the gas is determined experimentally. Hence the value of the mean free path of a molecule can be calculated from equation (i).

5.26 TRANSPORT PHENOMENA

According to Maxwell's law of distribution of velocity

$$dN = 4\pi NA^3\ e^{-bc2}\ c^2dc \qquad ...(i)$$

But $$4\pi\ c^2dc = du\ dv\ dw$$

$$dN = NA^3\ e^{-bc2}\ du\ dv\ dw \qquad ...(ii)$$

Also $$c^2 = u^2 + v^2 + w^2$$

$\therefore$ $$dN = NA^3e^{-bc2}\ (u^2 + v^2 + w^2)\ du\ dv\ dw \qquad ...(iii)$$

Equation (iii) has to be modified in case the gas as a whole possesses mass motion.

Let u_0, v_0 and w_0 be the components of the mass velocity. Therefore, the actual velocity of a molecule consists of two parts:

(i) the mass velocity components u_0, v_0 and w_0.

(ii) the random thermal velocity components.

$$u - u_0,\ v_0 - v_0\ w_0 - w_0.$$

Corresponding to thermal motion without mass motion, similar quantities with mass motion are

$$U = u - u_0$$

$$V = v - v_0$$

and $$W = w - w_0.$$

From equation (iii), Maxwell's law of distribution of velocity can be written as

$$dN = NA^3e^{-b(U2 + V2 + W2)}\ dU\ dV\ dW. \qquad ...(iv)$$

Equation (iv) holds good only if u_0, v_0, w_0, T and N are constant throughout the gas.

If the gas u not in an equilibrium state, there are three possibilities occurring singly or jointly:

(1) The components of velocity, u_0, v_0 and w_0, may not have the same value in all parts of the gas. This will result in relative motion of the gas layers with respect to one another.

There is relative velocity between different layers of the gas. This gives rise to the phenomena of viscosity.

(2) The temperature of the gas may not be the same throughout. This results in the transference of thermal energy from regions of higher to lower temperature. This gives rise to the phenomenon of conduction.

(3) The number of molecules per cc *i.e.*, N, may not be the same throughout the volume of the gas. This results in the movement of molecules from regions of higher value of N to lower value of N. This gives rise to the phenomenon of diffusion.

From the above discussion it is clear that the transport of momentum, energy and mass represent viscosity, conduction and diffusion respectively. These are called *transport phenomena.* From the thermodynamic point of view, the transport phenomena are irreversible.

The transport phenomena occur due to the thermal agitation of the molecules. The molecules in a particular layer are associated with certain values of velocity components, temperature and molecular density. They tend to minimise the differences in u_0, v_0, w_0 T and N, The molecules actually possess large velocities whereas these phenomena of viscosity, conduction and diffusion are comparatively slow. This anomaly is explained on the fact that the molecules frequently collide. Therefore, the transport phenomena are basically governed by the mean free path *A of a molecule. A molecule moving through a free path λ of a actually transferring momentum energy and sum through a distance λ.*

The transport phenomena occur only in the non-equilibrium state of a gas.

5.27 VISCOSITY OF GASES

Consider a gas flowing along a horizontal surface. The layer in contact with the wall of the surface is at rest. The velocity of the layer increases with increase in distance from the fixed layer.

Viscosity is defined as the tangential force per unit area required to maintain a unit velocity gradient.

$$F = -\eta A \frac{du}{dx} \qquad ...(i)$$

Here η is the coefficient of viscosity and A is the area of the layer and $\frac{du}{dx}$ is the velocity gradient.

The velocity of a molecule at any layer at a height x above the fixed layer $= x \times \frac{du}{dx}$. If the mean free path of a molecule is λ, then the distance for the first collision to occur $= (x - \lambda)$. The average velocity of these molecules $= (x - \lambda)\frac{du}{dx}$.

The molecules are moving at random in all directions. Approximately, it can be assumed that about one-third of the molecules are moving along each axis. As the molecules in a given space will be free to move in any direction, one-sixth- can be assumed to be moving in one direction and the other one-sixth in the opposite direction.

Let C be the r.m.s. velocity of the molecules, n the number of molecules per cc and m the mass of each molecule. Then, the number of molecules crossing unit area in one direction in one second = 1/6nC.

Mass of the molecules $= \frac{1}{6}\, mnC = M$

Momentum in the forward direction

$$= M\left[(x - \lambda)\,\frac{du}{dx}\right]$$

Momentum in the opposite direction

$$= M\left[\{(x - (-\lambda)\}\,\frac{du}{dx}\right]$$

Total change in momentum in one second

$$F = M(x - \lambda)\,\frac{du}{dx} - M\,(x + \lambda)\,\frac{du}{dx}$$

$$F = -2M\lambda\,\frac{du}{dx}$$

$$F = -2\left[\frac{1}{6}\,mnC\right]\lambda\,\frac{du}{dx}$$

$$= -\frac{1}{3}\,mnC\lambda\,\frac{du}{dx} \qquad ...(ii)$$

Comparing (i) and (ii)

$$\eta A = \frac{1}{3}\,mnC\lambda\,\frac{du}{dx}$$

But $\quad A = 1$ sq cm

$$\therefore \qquad \eta = \frac{1}{3} mnC\lambda \qquad ...(iii)$$

But
$$\lambda = \frac{1}{\sqrt{2}\,\pi\sigma^2 n}$$

$$\eta = \frac{1}{3} mnC \frac{1}{\sqrt{2}\,\pi\sigma^2 n}$$

$$\eta = \frac{mC}{3\sqrt{2}\,\pi\sigma^2} \qquad ...(iv)$$

But $\qquad C \propto \sqrt{}\,\sqrt{T}$

$\therefore \qquad \eta \propto \sqrt{T}.$

The coefficient of viscosity of a gas is directly proportional to the square root of its temperature in K.

5.28 THERMAL CONDUCTIVITY OF GASES

The thermal conductivity of gases can be explained on the basis of kinetic theory of gases. The gas molecules are free to move between the hot end and the cold end. They are also free to move in any direction. Let the temperature gradient be $\frac{d\theta}{dx}$ and the mean free path A. The temperature decreases from the hot to the cold end and when the steady state is reached, the temperature gradient will be uniform.

The mean temperature of the molecules crossing the layer P from the cold ride $= \left(\theta - \frac{d\theta}{dx}\lambda\right)$ and the mean temperature of the molecules crossing the layer P from the hot end $= \left(\theta + \frac{d\theta}{dx}\lambda\right)$.

Let n be the number of molecules per cc, m the mass of each molecule and C the r.m.s. velocity of the molecules. On the average only 1/6 of the molecules will be moving from the hot to the cold end and 1/6 of the molecules will be moving from the cold to the hot end. Consider one sq cm area of cross-section at the layer P (Fig. 5-12).

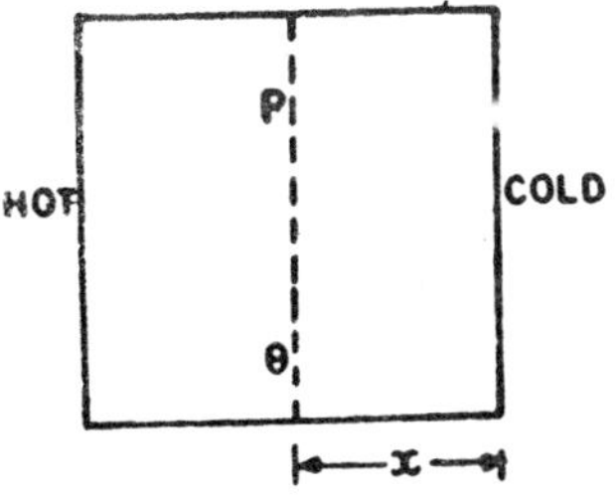

Fig. 5.12

The mass of the molecules crossing a unit area of cross-section in one second

$$= \frac{1}{6} mnC = M$$

$\therefore$ Heat energy carried by the molecules crossing the of the layer in; one second from the hot to the cold end

$$= MC_v\left(\theta + \frac{d\theta}{dx}\lambda\right)$$

Heat carried by the molecules crossing the area of the layer in one second from the cold to the hot end

$$= MC_v\left(\theta - \frac{d\theta}{dx}\lambda\right)$$

Therefore, the net heat conducted per unit area per second from the hot to the cold end

$$H = MC_V\left[\theta + \frac{d\theta}{dx}\lambda\right] - MC_V\left[\theta - \frac{d\theta}{dx}\lambda\right]$$

$$H = 2MC_V\lambda\frac{d\theta}{dx} \qquad \text{...(i)}$$

$$\therefore \qquad H = 2\left[\frac{1}{6} mnC\right]C_V\lambda\frac{d\theta}{dx}$$

$$H = \frac{1}{3} mnC.C_V\lambda\frac{d\theta}{dx} \qquad \text{...(ii)}$$

From the definition of thermal conductivity

$$H = KA.\frac{d\theta}{dx}.t$$

Here $A = 1$ sq cm

$t = 1$ second

$$\therefore \qquad H = K.\frac{d\theta}{dx} \qquad \text{...(iii)}$$

Comparing (ii) and (iii)

$$K = \frac{1}{3} mnC.C_V.\lambda \qquad \text{....(iv)}$$

$$K = \frac{1}{3} mnC.C, \times \frac{1}{\sqrt{2}.\pi\sigma^2 n}$$

$$= \frac{mC.C_V}{3\sqrt{2}.\pi\sigma^2} \qquad ...(v)$$

But $\qquad C \propto \sqrt{\sqrt{T}}$

$\therefore \qquad K \propto \sqrt{T}.$

The coefficient of thermal conductivity of a gas is directly proportional to the square root of the absolute temperature.

5.29 ATOMIC HEAT OF SOLIDS

A solid u made up of a number of atoms. The atoms can be considered as elastic spheres vibrating about their mean positions. The atoms will have three degrees of freedom in a solid. The kinetic energy for each degree of freedom = 1/2 kT. But an atom has potential energy also. The mean potential energy is equal to the mean kinetic energy. Therefore, the total energy associated for each degree of freedom of an atom = 2 × 1/2 kT. The total energy associated for three degrees of freedom = 3kT. The energy associated with 1 gram molecule

$$= N \times 3kT = 3RT$$

$$U = 3RT$$

$$A_H = \frac{dU}{dT}$$

$$= 3R = 5.96 \text{ cals/g-mole K.}$$

Thus, the atomic heat of solids (Ay) is 35 and this value agrees with the Dulong and Petit's Law.

5.30 CHANGE OF STATE

The molecules in a liquid have random motion. The average kinetic energy of the molecules depends upon the temperature. The molecules are held together to occupy a particular volume due to intermolecular attraction.

Consider water kept in a tray and exposed to the atmosphere. Generally, the atmosphere is not saturated with water vapours. The molecules, of the liquid which are in a state of constant random motion collide with each other resulting in exchange of kinetic energies. Thus, there exists the possibility of a particular molecule in the vicinity of water, air interface, acquiring sufficiently high kinetic energy. This energy of the molecule may be high enough to enable the molecule to break away from the forces of attraction of the surrounding molecules. In fact, a molecule on the interface experiences minimum force of attraction. Such a molecule flies off into vapour space. When this molecule leaves the liquid, the mean kinetic energy of the remaining liquid molecules decreases and there is fall in temperature. This explains cooling caused by evaporation. Some vapour molecules in the atmosphere, also coalesce and get absorbed by the water surface. This is called condensation. Evaporation and condensation continue simultaneously but rate of evaporation is higher than the rate of condensation.

An essential precondition for cooling due to evaporation if the capacity of air to absorb more moisture.

If the liquid is heated, and if the rise in temperature of the liquid is more than the cooling caused by evaporation, the temperature of the liquid increases in addition to the increase in the rate of evaporation. A stage will come when the rate of heat supply is exactly equal to the energy needed for evaporation. At this stage evaporation takes place without change in temperature. This is boiling. At the boiling point the molecules are pulled further apart and their positions are no longer restricted to the space occupied by the liquid. This explains the change of state from liquid to gas or Vapour. The amount of work done in pulling the molecules apart measures the latent heat of vaporisation. At the boiling point, the heat supplied to the liquid is used up only in pulling the molecules apart against the forces of intermolecular attraction and hence no rise in temperature of the liquid is noticed till the whole of the liquid is changed to the gaseous state. 1 cc of water when converted to steam occupies a volume of 1600cc. This means that the mean intermolecular distance is approximately 11.7 times more ($\sqrt[3]{1600}$ = 11.7 approximately) in a gas than the corresponding liquid.

5.31 CONTINUITY OF STATE

A substance can exist in three states viz., solid, liquid and gas. For a long time it was supposed that gases like hydrogen, nitrogen, oxygen, helium etc. exist only in the gaseous form. It was also supposed that liquid and gaseous states are not continuous but are distinctly separate from one another. It was found that at high pressures and low temperatures, there are marked deviations from Boyle's law. At high pressure, the molecules of a gas come close to each other and the forces of intermolecular attraction become appreciable. If these molecular distances become equal to those of the corresponding liquid, the gas gets liquefied.

The three states of a substance are continuous *i.e.*, solid, liquid and gaseous states are the three different stages of a continuous phenomenon. If ice is heated, its molecules become more free to move and it is converted to the liquid state (water). Further heating increases the molecular movement and when intermolecular forces become small, at a particular stage it is converted into the gaseous; state (steam). The reverse process is also possible by continuously withdrawing heat.

Faraday was able to liquefy chlorine under its own pressure with the help of a freezing mixture.

Cagniard de la Tour's Experiment

The apparatus consists of a tube having a bulb B. It contains mercury and above the surface of mercury in B, the bulb contains water or alcohol (Fig. 5.13). There is air at the top end of the tube A.

The space above the liquid in the bulb B contains only the saturated vapours of the liquid at room temperature. The free surface of the liquid is visible and there is a marked distinction in the boundary of the liquid and its vapour. The bulb B is gradually heated. After some time the whole space in the bulb B appears foggy though the whole of the liquid has not evaporated. There is no free surface of the liquid at this stage. This shows that there is no distinction in the boundary of the liquid and the gaseous states.

When the bulb is allowed to cool, a cloud appeals and it gets condensed. Again the free surface of the liquid is visible. This shows that there is a continuity between the liquid and the gaseous states of a substance. They are two different stages of a continuous phenomenon.

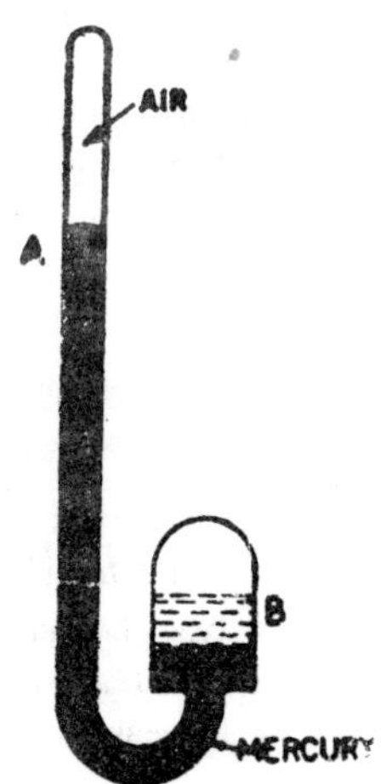

Fig. 5.13

5.32 ANDREWS' EXPERIMENTS ON CARBON DIOXIDE

The apparatus used by Andrews to study the isothermal of carbon dioxide at various temperatures is shown in Fig. 5.14. A and B are two similar glass tubes having thick capillary tubes at the top and bulbs in the middle. Initially in the tube A pure dry air is passed for a long time and it is sealed. In the bulb B carbon dioxide is passed for a long time (say 24 hours) and the ends are sealed. The lower ends of both the tubes are immersed in mercury and the lower ends of the tubes are opened under mercury. A small pellet of mercury is drawn in both the tubes by alternately heating and cooling the tubes. These mercury pellets act as stoppers. Both the tubes are fixed in a H-shaded copper vessel having steel stoppers S, S.

The vessel contains, water and the screws are turned so that the mercury pellets in the two tubes appear above the vessel. The capillary tubes of A, and B are calibrated to read volume directly.

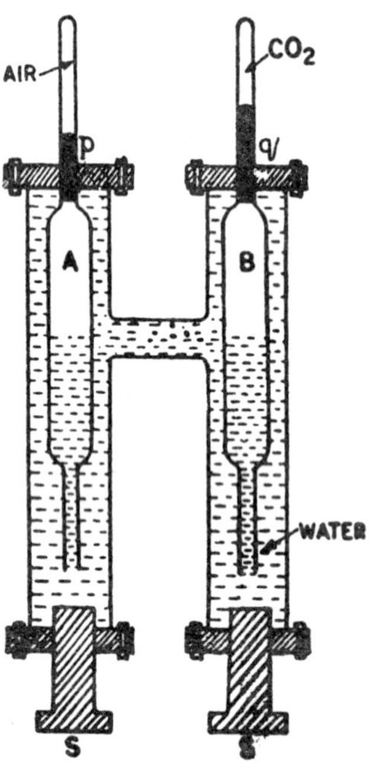

Fig. 5.14

Since the pressures of air in A and CO, in B are the same, from the volume of air in A, pressure of CO_2 in B can be calculated

The volume of CO_2 is read directly from the tube B.

The experiment is performed by maintaining the capillary tube of A at a fixed temperature and the capillary tube of B at any desired temperature. Different pressures are applied at a fixed temperature of CO_2 and the readings are taken until the volumes are too small. Pressure can be increased by screwing in the plungers S. S.

The experiment is repeated, maintaining CO_2 at different temperatures. Thus different sets of readings of pressure and volume for carbon dioxide at various temperatures are obtained. A graph is plotted between pressure and volume at various temperatures (Fig. 5.15).

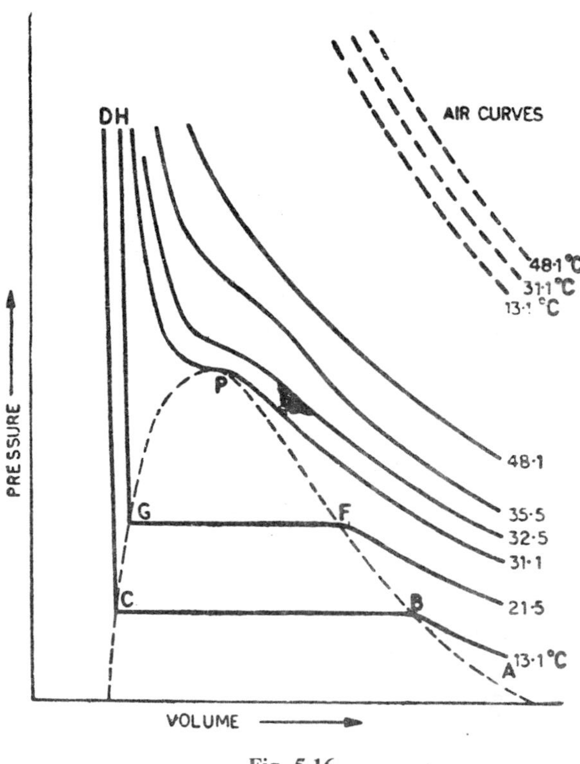

Fig. 5.16

The isothermals of CO_2 at various temperatures are shown in Fig. 5.15.

At 13.1°C, the portion AB represents the gaseous state of CO_2. Upto the point B, it obeys Boyle's law. From B to C it shows an enormous decrease in volume with flight increase in pressure. The portion BC represents the change of CO_2 from the gaseous to the liquid state. At C the whole of the gas has been liquefied. The portion CD represents the liquid state of CO_2 *i.e.*, there is no appreciable decrease in volume with increase in pressure.

At 21.5°C the curve is similar. The horizontal portion of the curve (FG) has decreased.

At 31.1°C the horizontal portion vanishes. It shows that the gas cannot be liquefied any more and its volume diminishes with increase in pressure.

The isothermals at 32.5°C and 35.5°C show the same behaviour and the portion of large compressibility decreases.

At 48.1°C the curve becomes smooth and it is similar to that of a permanent gas and it obeys Boyle's law.

Andrews found that the liquefaction of CO_2 occurred only below 31.1°C. This temperature is called the critical temperature.

Critical temperature of a gas is defined as that temperature below which the gas can be liquefied by the application of pressure alone. Above the critical temperature, the gas cannot be liquefied howsoever large the applied pressure may be.

In the graph, the pressure, volume and temperature corresponding to the point P are called the critical pressure (P_e), critical volume (V_e) and critical temperature (P_e).

Critical pressure is the pressure applied to the gas at its critical temperature so that it gets liquefied.

Critical volume is the particular volume of a gas at critical pressure and critical temperature.

Results

(i) There is no physical distinction between the liquid and the gaseous state of a substance. This is shown by the horizontal portions of the isothermal.

(ii) Below the critical temperature a gas can be liquefied under the application of pressure alone. The critical temperature for oxygen is –118.8°C; for nitrogen 147.1°C; and for hydrogen –240°C. Even these gases can be liquefied under the application of pressure if they are below the critical temperature. These gases have also been liquefied and they are no longer permanent gases.

(iii) The density of the vapour and the liquid became equal at the critical point. At the critical temperature the boundary line between the vapour and the liquid vanishes. Moreover, the compressibility of the vapour is infinite at the critical point.

5.33 AMAGAT'S EXPERIMENT

Amagat performed a series of experiments with hydrogen (1878) and nitrogen (1893) to study their behaviour at high pressures. The apparatus consisted of a large metallic cylinder containing mercury having a screw S fitted to its one end. The apparatus was arranged at the base of a mine. A steel manometer tube of length about 300 metres was fixed along the shaft of the coal mine. The manometer tube was open at the top end. A calibrated inverted tube A containing nitrogen (or hydrogen) was fitted to the cylinder containing mercury (Fig. 5.16). An oil bath surrounded the tube A and the temperature of nitrogen could be kept fixed at any desired value. The tube A was calibrated to read the volume of the gas directly.

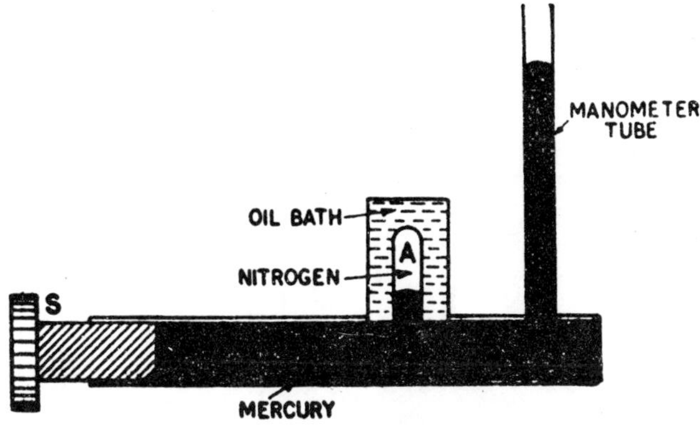

Fig. 5.16

The difference in levels in mercury columns in the tube A and the manometer tube measured the pressure on the gas. With this arrangement pressures up to 420 atmospheres could be applied.

Amagat plotted the isothermals between PV and P from the experimental data. He obtained the following results:

(1) For gases like air, oxygen, nitrogen, CO_2, ethylene etc., at low pressure, the product PV decreases with increase/in P.

(2) With increase in pressure, the product PV attains a minimum value and then shows a steady increase with increase in P.

(3) In the case of hydrogen, the value of PV increases steadily with increase in P.

5.34 HOLBORN'S EXPERIMENT

In 1915, Holborn and his co-workers performed a series of experiments to study the behaviour of gases at high pleasures.

The apparatus consists of a glass bulb A connected to a manometer M, the globes B_1, B_2, B_3 ..., etc., and the pressure, balance (Fig. 5.17).

The pressure balance consists of a fixed block E. A piston C fits into the hole in the block S. The piston C balances the pressure exerted by the gas through the oil in (Be cavity of the block. The piston C is kept at a fixed mark by adjusting the weight W. The total weight exerted on the oil is due to the weight of W, D, S, and C. Let the total weight be Mg and the area of cross-section of the piston C be a. The downward, pressure due to the weight = Mg/a This is also equal to the pressure of the gas in the bulb A.

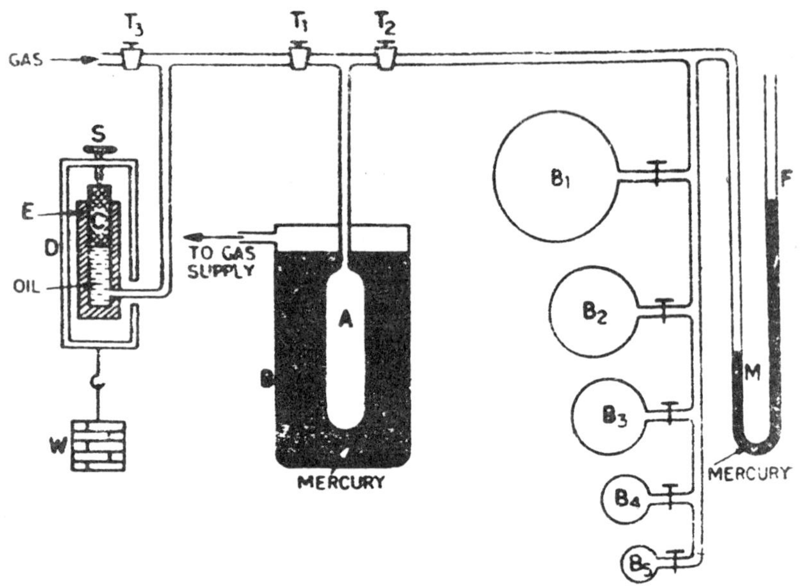

Fig. 5.17

Initially the bulbs B_1, B_2 and B_3... etc., and A are evacuated. The stop-cock T_2 is closed. Gas at high pressure is admitted by opening the stop-cocks T_3 and T_1. The stop-cock T_3 is closed and the pressure of the gas in A is measured with the help of the pressure balance as explained above.

The stop-cock T_1 is closed and the stop-cock T_2 is opened. The stop-cock of the bulbs B_1, B_2 etc., are opened one after the other till the pressure of the gas in A and the bulb is approximately equal to the atmospheric pressure. The volume of the bulbs, the connecting tubes and the bulb A is accurately known. The bulb A is enclosed in a bath containing the gas at high pressure to avoid the breakage of the bulb at high pressure.

Let the pressure of the gas as measured with the pressure balance be P_1, where $P_1 = Mg/a$. Let m_1 be the mass of the gas enclosed in A and m be the gram molecular weight of the gas. Then.

$$PV_1 = \frac{m_1}{m}. RT$$

Here V_1 is the volume of the gas after expansion.

From this, the value of m_1 is calculated. Here P is the pressure of the gas after expansion and T is the temperature of the gas. If the volume of the gas admitted initially in the bulb A = V_e then m_1 grams of the gas at a pressure P_1 occupies volume V_e at a temperature T.

Suppose in the second set of readings P_2 is the pressure of the gas and V_2 is the volume after expansion. If m_2 is the mass of the gas in A, then

$$PV_2 = \frac{m_2}{m} RT$$

From this equation m_2 is calculated.

If V_0 Is the volume of the gas in A before expansion, then the volume of the gas for m_1 grams at a pressure P_2

$$= \frac{V_0 \times m_2}{m_2}$$

Thus, the volume occupied by the same mass of the gas for a range of pressures and at different temperatures can be calculated.

5.35 BEHAVIOUR OF GASES AT HIGH PRESSURE

The isothermals between PV and P for various gases have been drawn and the general nature of the curves is as shown in Fig. 5.18. The general nature of the curves is the same for all gases.

(1) At high temperature, the value of PV increases with increase in P.

(2) At lower temperatures, the value of PV decreases initially with increase in P. It becomes a minimum at a particular pressure and

then increases with increase in pressure. The locus of these minima is shown by the dotted curve A.

(3) At temperatures below the critical temperature, there is a sudden decrease in the value of PF with increase in pressure. This corresponds to the change of state from gas to liquid. When the liquefaction is complete, PV gradually increases with increase in P, The shaded area represents the region of liquefaction

Based on the experimental results of Andrews, Amagat and Holborn, K. Onnes suggested an empirical relation

$$PF = A + BP + CP^2 + DP^3 \ldots.. \qquad \ldots(i)$$

Here A, B, C, D etc., are constants depending on the nature of the gas. These constants are called *virial coefficients*. These constants change with temperature. The constant B has a special importance. Its value is negative at low temperatures and if value increases through zero to positive values at higher temperatures. The temperature at which the coefficient B is zero is called the *Boyle Temperature*. At the Boyle temperature,

$$PV = A = \text{constant}$$

and $$B = \frac{d[PV]}{dP} = 0$$

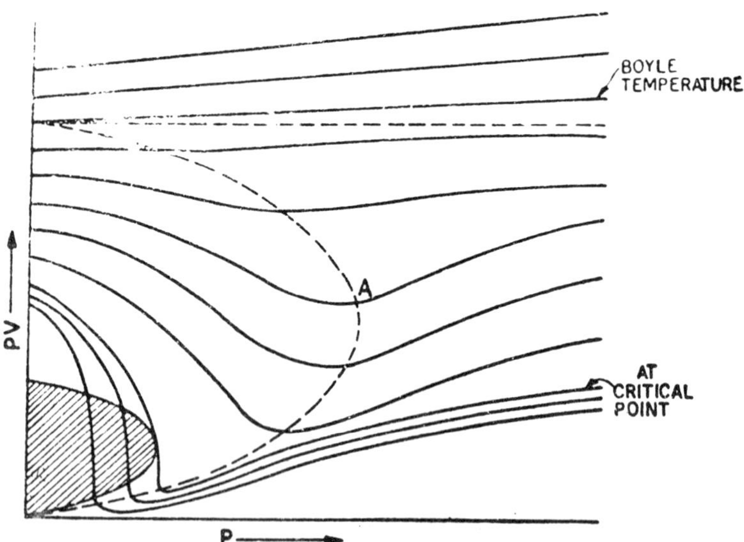

Fig. 5.18

Below the Boyle temperature, the gases are highly compressible and this suggests the existence of intermolecular attractions. Beyond the Boyle temperature, Boyle's law is obeyed and intermolecular attractions are less significant.

5.36 VAN DER WAALS EQUATION OF STATE

While deriving the prefect gas equation PV = RT on the basis of kinetic theory, it was assumed that (i) the size of the molecule of the gas is negligible and (ii) the forces of intermolecular attraction are absent. But in actual practice, at high pressure, the size of the comparison with the volume of the gas. Also, at high pressure, the molecules come closer and the forces of intermolecular attraction are appreciable. Therefore, correction should be applied to the gas equation.

(i) Correction for Pressure

A molecule in the interior of a gas experiences forces of attraction in all directions and the resultant cohesive force is zero. A molecule near the walls of the container experiences a resultant force inwards (away from the wall). Due to this reason the observed pressure of the gas is less than the actual pressure. The correction for pressure p depends upon (i) the number of molecules striking unit area of the walls of the container per second and (ii) the number of molecules present in a given volume. Both these factors depend on the density of the gas.

$\therefore$ Correction for pressure $p \propto \rho^2 \propto \frac{1}{V^2}$

$$p = \frac{a}{V^2}$$

Here a is a constant and V is the volume of the gas

Hence correct pressure

$$(P + P) = P + \frac{a}{V^2}$$

where P is the observed pressure.

(ii) Correction for Volume

The fact that the molecules have finite size shows that the actual space for the movement of the molecules is less than the volume of the vessel. The molecules have the sphere of influence around them and due to this factor, the correction for volume is b where b is approximately four times the actual volume of the molecules. Therefore the corrected volume of the gas = (V – b).

Let the radius of one molecule be r.

The volume of the molecule $= x = \frac{4}{3}\pi r^3$

The centre of any two molecules can approach each other only by a minimum distance of 2r *i.e.*, the diameter of each molecule. The volume of the sphere of influence of each molecule,

$$S = \frac{4}{3}\pi(2r)^3 = 8x.$$

Consider a container of volume V. If the molecules axe allowed to enter one by one,

The volume available for first molecule = V

Volume available for second molecule = V – S

Volume available for third molecule = V – 2S

....

Volume available for nth molecule = V – (n – 1)S

Average space available for each molecule

$$= \frac{V + (V - S) + (V - 2S) + \ldots\ldots \{V - (n-1)S\}}{n}$$

$$= V - \frac{S}{n}\{1 + 2 + 3 \ldots. (n-1)\}$$

$$= V - \frac{S}{n} \cdot \frac{(n-1)\,n}{2}$$

$$= V - \frac{nS}{2} + \frac{S}{2}$$

As the number of molecules is very large S/2 can be neglected.

∴ Average space available for each molecule

$$= V - \frac{nS}{2} \quad \text{(But } S = 8x\text{)}$$

$$= V - \frac{n(8x)}{2}$$

$$= V - 4(nx)$$

$$= V - b$$

$\therefore \quad b = 4(nx) =$ four times the actual volume of the molecules

Thus the Van der Waals equation of state for a gas is

$$\left(P + \frac{a}{V^2}\right)(V - b) = RT \qquad \text{...(i)}$$

where a and b are Van der Waals constants.

From the Van der Waals equation of state

$$\left(P + \frac{a}{V^2}\right)(V - b) = RT$$

$$P = \frac{RT}{V - b} - \frac{a}{V^2} \qquad \text{...(ii)}$$

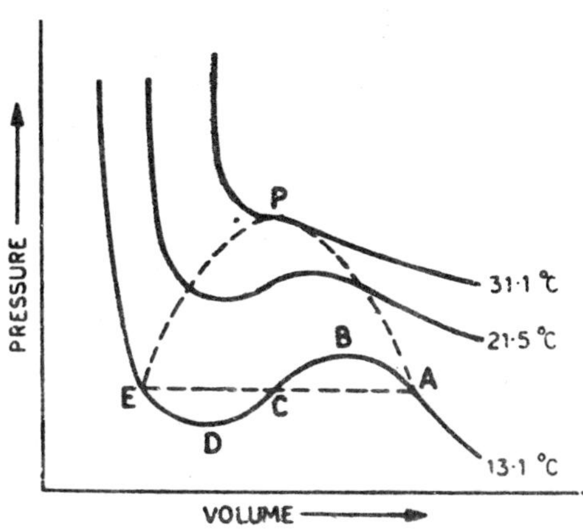

Fig. 5.19

Graphs between pressure and volume at various temperatures are drawn using equation (i). The graphs are as shown in Fig. 5.19. In the graph, the horizontal portion is absent. But in its place, the curve ABCDE is obtained. This does not agree with the experimental isothermals for CO_2 as obtained by Andrews. However, the portion AB has been explained as due to supercooling of the vapours and the portion ED due to superheating of the liquid. But the portion BCD cannot be explained became it shows decrease in volume with decrease in pressure. It is not possible in actual practice. The states AS and ED, though unstable, can be realised in practice by careful experimentation. At higher temperatures, the theoretical and experimental isothermals are similar.

Until now as many as 56 different equations of state have been suggested. But no single equation satisfies all the observed facts.

Dicterice (1901) has suggested an equation

$$P(V-b) = RTe^{-\frac{a}{RTV}}$$

Berthelot has suggested an equation

$$\left(P + \frac{a}{V^2T}\right)(V-b) = RT.$$

5.37 CRITICAL CONSTANTS

The critical temperature and the corresponding values of pressure and volume at the critical point are called the critical constants. At the critical point, the rate of change of pressure with volume (dP/dV) is zero. This point is called the point of inflexion.

According to Van der Waals equation

$$\left(P + \frac{a}{V^2}\right)(V-b) = RT \quad \text{...(i)}$$

$$P = \left(\frac{RT}{V-b}\right) - \frac{a}{V^2} \quad \text{...(ii)}$$

Differentiating P with respect to V

$$\frac{dP}{dV} = \frac{-RT}{(V-b)^2} + \frac{2a}{V^3} \quad \text{...(iii)}$$

At the critical point dP/dV = 0

$$T = T_e$$

$$V = V_e$$

$$\therefore \quad \frac{-RT_e}{(V_e-b)^2} + \frac{2a}{V_e^3} = 0$$

or

$$\frac{2a}{V_e^3} = \frac{RT_e}{(V_e-b)^2} \quad \text{....(iv)}$$

Differentiating equation (iii)

$$\frac{d^2P}{dV^2} = \frac{2RT}{(V-b)^3} - \frac{6a}{V^4}$$

At the critical point $\frac{d^2P}{dV^2} = 0,$

$$T = T_e$$

$$V = V_e$$

$$\therefore \qquad \frac{2RT_e}{(V_e - b)^3} - \frac{6a}{V_e^4} = 0$$

$$\frac{6a}{V_e^4} = \frac{2RT_e}{(V_e - b)^3} \qquad \text{...(v)}$$

Dividing (iv) by (v)

$$\frac{V_e}{3} = \frac{V_e - b}{2}$$

or

$$2V_e = 3V_e - 3b$$

$$V_e = 3b \qquad \text{...(vi)}$$

Substituting the value of

$V_e = 3b$ in equation (iv)

$$\frac{2a}{27b^2} = \frac{RT_e}{4b^2}$$

or

$$T_e = \frac{8a}{27Rb} \qquad \text{...(vii)}$$

Substituting these values of V_e and T_e in equation (ii)

$$P_e = \frac{R \times 8a}{27Rb(2b)} - \frac{a}{9b^2}$$

$$P_e = \frac{a}{27b^2} \qquad \text{...(viii)}$$

5.38 CORRESPONDING STATES

Two gases are said to be in corresponding states if the ratios of their actual pressure, volume and temperature and critical pressure, critical volume and critical temperature have the same value. It means

$$\frac{P_1}{P_{e1}} = \frac{P_2}{P_{e2}}.$$

$$\frac{V_1}{V_{e1}} = \frac{V_2}{V_{e2}}$$

and $$\frac{T_1}{T_{e1}} = \frac{T_2}{T_{e2}}$$

Critical Temperature and Pressure of Common Gases

Substance	Critical Temperature °C	Critical Pressure (Atmospheres)
Helium	–268°C	2.26
Hydrogen	–240°C	12.80
Nitrogen	–146°C	33.50
Air	–140°C	39.00
Argon	–122°C	48.00
Oxygen	–118°C	50.00
Carbon dioxide	31°C	73.00
Ammonia	130°C	15.00
Chlorine	146°C	76.00
Water	374°C	218.00

5.39 COEFFICIENT OF VAN DER WAALS CONSTANTS

$$V_e = 3b \qquad \text{...(i)}$$

$$P_e = \frac{a}{27b^2} \qquad \text{...(ii)}$$

$$T_e = \frac{8a}{27\ Rb} \qquad \text{...(iii)}$$

The equations (iii) and (ii)

$$\frac{T_e^2}{P_e} = \frac{64\,a^2}{(27)^2\ R^2b^2} \times \frac{27\,b^2}{a}$$

$$= \frac{64\,a}{(27)\ R^2}$$

$$a = \frac{27}{64}\frac{R^2T_e^2}{P_e} \qquad \text{...(iv)}$$

Dividing (iii) by (ii)

$$\frac{T_e}{P_e} = \frac{8a^2}{27\,Rb} \times \frac{27b^2}{a}$$

$$= \frac{8B}{R}$$

$$b = \frac{RT_e}{8P_e} \qquad \text{...(v)}$$

$$\frac{RT_e}{P_eV_e} = \frac{R\,(8a)\,.\,27b^2}{27\,Rba\,(3b)}$$

$$\frac{RT_e}{P_eV_e} = \frac{8}{3} \qquad \text{...(vi)}$$

The quantity $\frac{RT_e}{P_eV_e}$ is called the *critical coefficient* of a gas.

Its calculated value = 8/3 and it is the same for all gases.

The experimental values of the critical coefficient for different gases is given below.

Experimental Values of Critical Coefficient $\frac{RT_e}{P_eV_e}$.

Substance	T_e in K	P_e in atm.	Specific volume in cm^3/g	RT_e/P_eV_e
Helium	51	2.25	15.4	3.13
Hydrogen	33.1	12.8	32.2	3.28
Nitrogen	125.9	33.5	3.21	3.42
Oxygen	154.2	49.7	2.32	3.42
Carbon dioxide	304	72.8	2.17	3.48
Water	647	218	3.181	4.30

Note: Here V_e = molecular wt × specific volume.

The experimental value of the critical coefficient of all gases is greater than the theoretical value of 2.67.

5.40 REDUCED EQUATION OF STATE

Taking the pressure, volume and temperature of a gas in terms of reduced pressure, volume and temperature,

$$\frac{P}{P_e} = \alpha, \frac{V}{V_e} = \beta, \frac{T}{T_e} = \gamma$$

$$P = \alpha\, P_e,\ V = \beta\, V_e,\ T = \gamma\, T_e$$

According to Van der Waals equation

$$\left(P + \frac{a}{V^2}\right)(V - b) = RT \qquad \text{...(i)}$$

or $$\left(\alpha P_e + \frac{a}{\beta^2 V_e^2}\right)(\beta\, V_e - b) = R\gamma\ T_e$$

But $$P_e = \frac{a}{27\,b^2}$$

$$V_e = 3b$$

and $$T_e = \frac{8a}{27\,Rb}$$

$$\therefore \quad \left[\frac{\alpha . a}{27\,b^2} + \frac{a}{\beta^2\, 9b^2}\right][\beta . 3b - b] = \frac{R\gamma\ 8a}{27\ Rb}$$

$$\left[\alpha + \frac{3}{\beta^2}\right][3\beta - 1] = 8\gamma \qquad \text{...(ii)}$$

This is the reduced equation of state for a gas. If two gases have the same values of α, β and γ they are said to be in corresponding states.

5.41 PROPERTIES OF MATTER NEAR CRITICAL POINT

Based on the experiments of Andrews, Amagat and others, the state of matter near the critical point can be summarised as follows:

(1) The densities of the vapour and the liquid gradually approach each other and their densities are equal at the critical point.

(2) At the critical point or just near the critical point, the line of demarcation between the liquid and the vapour disappears. Consequently, there must exist mutual diffusion between the two phases and the surface tension must be zero. This also

means that the forces of intermolecular attraction in the liquid and vapour states must be equal at the critical point.

(3) The whole mass of the substance exhibits a flickering experience which indicates that there may be variations of density inside the mass. This has been experimentally verified.

(4) The compressibility of the vapour becomes infinite at the critical temperature.

The experiments performed by Callendar and others have shown that the densities of the liquid and vapour are not equal at the critical temperature. The densities are equal at a temperature (lightly higher than the critical temperature. The critical temperature of water is 374°C. It was found that the densities are equal at 380°C. At 374°C the latent heat of water was found to be 72.4 calories per gram. At 380°C the latent heat of water is zero.

To conclude, the critical point for any substance is not a particular temperature but it is a narrow region (critical region). The critical point is that point at which the properties of the two phases become identical.

5.42 EXPERIMENTAL DETERMINATION OF CRITICAL CONSTANTS

(1) The critical constants can be determined by plotting the isothermals between P and V at different temperatures of the gas. The critical point is found on the graph, where the horizontal portion of the curve (change from gaseous to the liquid state) just vanishes. The pressure, volume and temperature corresponding to the critical point give the *critical constants* of the gas.

(2) With the help of Cagniard de la Tour's apparatus, the temperatures at which the liquid meniscus disappears and reappears are carefully noted. The mean of these two temperatures gives the critical temperature of the gas. The bulb is surrounded by a bath whose temperature is controlled by a sensitive thermostat and the temperature is varied by small amounts (say 0.1°C). The pressure is read with the help of a manometer. The pressure corresponding to the critical temperature gives the critical pressure. However, the measurement of critical volume offers a practical difficulty because even if the temperature is

slightly below the critical temperature, the change in volume is enormous even for a small change in pressure.

(3) Cailetet and Mathias have suggested a method for the determination of critical volume. Density of the liquid is determined accurately at different temperatures. The density of the saturated vapour is also determined accurately for the same temperature range. A graph is plotted between density and temperature (Fig. 5.20).

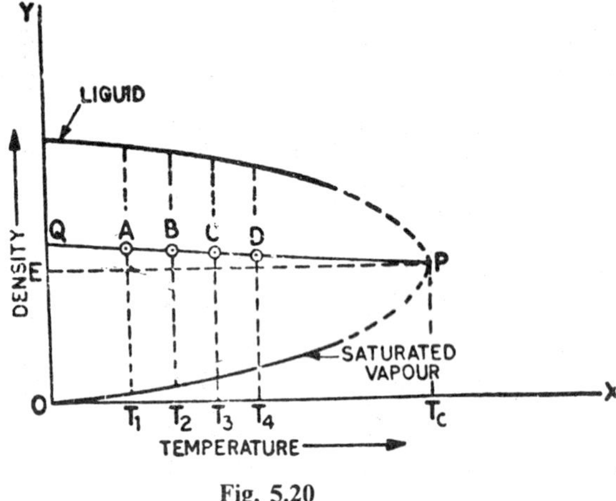

Fig. 5.20

The density of the liquid decreases with rise in temperature and that of the saturated vapour increases with rise in temperature. On the graph the points A, B, C and D represent the mean density of liquid and the vapour at temperatures T_1, T_2, T_3, T_4 etc., respectively.

The points lie on a straight line. The line is produced and the curves are extrapolated to meet the line at the point P. The density corresponding to the point P is obtained from the graph. Let this density be p.

The specific volume *i.e.*, volume per unit mass

$$V_e = \frac{1}{\rho}$$

This represents the critical volume. The temperature corresponding to the point P gives the critical temperature.

5.43 INTERMOLECULAR ATTRACTION

Joule's Experiment

Joule performed an experiment to study the intermolecular attraction between the molecules of a gas. He designed an apparatus to measure the change in energy of a gas when its pressure and volume are changed.

A and B are two flasks connected by a tube and fitted with a stop-cock 5. A was filled with the gas at high pressure and B was perfectly evacuated (Fig. 5.21). The bulbs were surrounded by a water bath and its temperature was noted with a sensitive thermometer. The stop-cock was opened and the gas rushed from A to B. If work was done during expansion, cooling was expected.

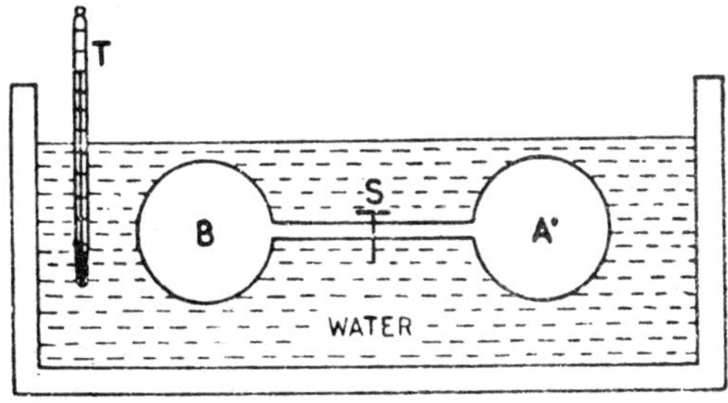

Fig. 5.21

But in this experiment no cooling was observed. This was puzzling as the work had been done by the gas in overcoming the forces of intermolecular attractions. Joule was able to explain the drawback in his experiment by assuming that the gas has expanded into vacuum and so no cooling was observed.

In this experiment only the pressure and volume of the gas changed and as the bath did not show any change in temperature, it is evident that the temperature of the gas remained constant.

According to Joule, this experiment also shows that the internal energy of a gas is a function of temperature only, irrespective of the changes in pressure and volume.

The porous plug experiment designed by Joule and Kelvin, however, showed cooling due to the work done by a gas in overcoming the intermolecular attraction when it was passed from the high pressure to the lov/ pressure side.

5.44 POROUS PLUG EXPERIMENT

The apparatus consists of a porous plug having two perforated brass discs D, D. The space between them is packed with cotton wool or silk fibres. The porous plug is fitted in a cylindrical box wood tube which is surrounded by a vessel containing cotton wool (Fig. 5.22). This is done to avoid loss or gain of heat from the surroundings. T_1 and T_2 are two sensitive platinum resistance thermometers and they measure the temperatures of the incoming and the outgoing gas. The gas is compressed to a high pressure with the help of the piston P and it is passed through a spiral tube immersed in a water bath maintained at a constant temperature. If there is any heating of the gas due to compression, this heat is taken by the circulating water in the water bath.

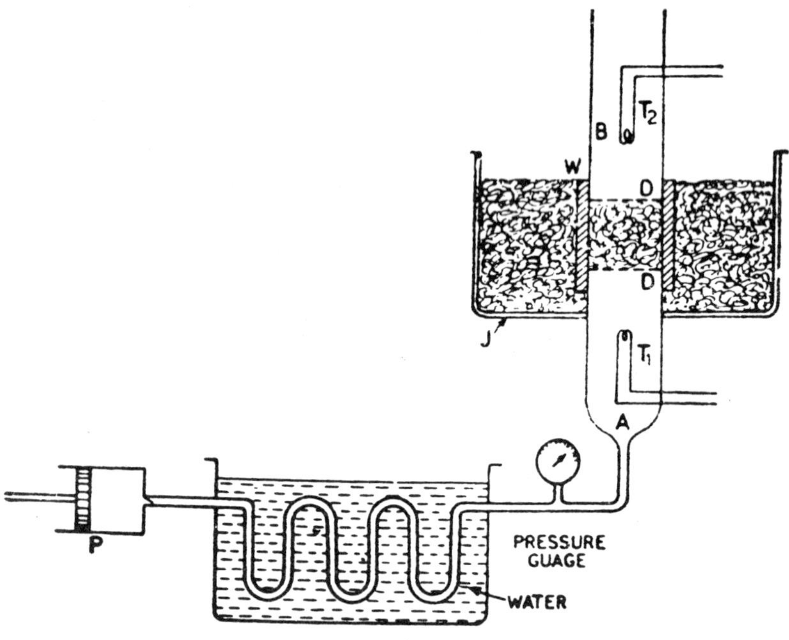

Fig. 5.22

The compressed gas is passed through the porous plug. The gas gets throttled (wire drawn) due to cotton wool. Work is done by the gas in overcoming intermolecular attraction. The temperature of the outgoing gas is measured with the help of a platinum resistance thermometer T_2. The pressure of the incoming gas is measured with the help of a pressure gauge and the pressure of the outgoing gas is equal to the atmospheric pressure.

The behaviour of a large number of gases was studied at various inlet temperatures of the gas and the results obtained are as follows:

(i) At sufficiently low temperatures all gases show a cooling effect.

(ii) At ordinary temperatures all gases except hydrogen and helium show cooling effect. Hydrogen shows heating instead of cooling at room temperature.

(iii) The fall in temperature is directly proportional to the difference in pressure on the two sides of the porous plug.

(iv) The fall in temperature per atmosphere difference of pressure decreases as the initial temperature of the gas is raised. It becomes zero at a particular temperature and at a temperature higher than this temperature, instead of cooling, heating is observed. This particular temperature at which the Joule-Thomson effect changes sign is called the temperature of inversion.

In the case of hydrogen heating was observed at room temperature because it was at a temperature far higher than its temperature of inversion. The temperature of inversion for hydrogen is –80°C and for helium it is –258°C. If helium is passed through the porous plug at a temperature lower than –258°C it will also show cooling effect. It means any gas below the temperature of inversion shows a cooling effect when it is passed through the porous plug or a throttle valve.

5.45 THEORY OF POROUS PLUG EXPERIMENT

The simple arrangement of the porous plug experiment is shown in Fig. 5.21. The gas is allowed to pass through the porous plug from the high pressure side to the low pressure side. The velocity of the gas as it emerges from the porous plug is very high and there is increase in the kinetic energy of the molecules. Consider one gram molecule of a gas to the left and to the right of the porous plug. Let P_1, V_1 and P_2, V_2 represent the pressure and volume on the two sides of the porous plug. When the piston A is moved through a certain distance dx, the piston B also moves through the same distance dx. The work done on the gas by the piston A = $P_1\ A_i\ dx = P_1\ V_1$. The work done by the gas on the piston B is $P_1A_2\ dx = P_2V_2$. Thus, the net external work done by the gas is $P_2V_2 - P_1V_1$.

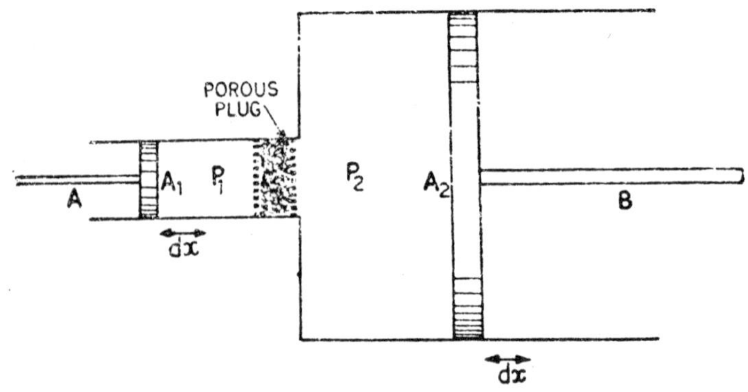

Fig. 5.23

If w is the work done by the gas in separating the molecules against their intermolecular attractions, the total amount of work done by the gas is

$$(P_2V_2 - P_1V_1) + w$$

No heat is gained or lost to the surroundings. There are three possible cases t

(i) Below the Boyle temperature,

$$P_1V_1 < P_2V_2$$

and $P_2V_2 - P_1V_1$ is +ve. w must be either positive or zero. Thus a net +ve work is done *by the gas* and there must be cooling when the gas passes through the porous plug.

(ii) At the Boyle temperature if P_1 is not very high

$$P_1V_1 = P_2V_2$$

and $\quad P_2V_2 - P_1V_1 = 0.$

The total work done by the gas in this case is w. Therefore cooling effect at this temperature is only due to the work done by the gas in overcoming intermolecular attractions.

(ii) Above the Boyle temperature

$$P_1V_1 > P_2V_2$$

or $\quad P_2V_2 - P_1V_1$ is –ve.

Thus, the observed effect will depend upon whether $(P_2V_2 - P_1V_1)$ is greater or less than w.

If $w > (P_1V_1 - P_2V_2)$, cooling will be observed.

If $w < (P_1V_1 - P_2V_2)$, heating will be observed.

Thus, the cooling or heating of a gas due to free expansion through a porous plug from a high pressure to a low pressure side will depend on (i) the deviation from Boyle's law and (ii) work done in overcoming intermolecular attractions.

5.46 JOULE-KELVIN EFFECT—TEMPERATURE OF INVERSION

Assuming that the Van der Waals equation is obeyed, the attractive forces between the molecules are equivalent to an internal pressure a/V^2.

When the gas expands from V_1 to V_2 the work done in overcoming intermolecular attractions

$$w = \int_{V_1}^{V_2} p.dV$$

But $$p = \frac{a}{V^2}$$

$\therefore$ $$p = \int_{V_1}^{V_2} \frac{a}{V^2} dV$$

$$= -\frac{a}{V^2} + \frac{a}{V_1}$$

If V_1 and V_2 represent the gram molecular volumes on the high and the low pressure sides respectively, the external work done by the gas is

$$(P_2V_2 - P_1V_1)$$

Hence the total work done by the gas

$$W = (P_2V_2 \;\; P_1V_1) + w$$

$$= (P_2V_2 - P_1V_1) - \frac{a}{V_2} + \frac{a}{V_1}$$

Van der Waals equation of state for a gas is

$$\left(P + \frac{a}{V^2}\right)(V - b) = RT$$

or $$PV + \frac{a}{V} - bP - \frac{ab}{V^2} = RT$$

or $$PV = RT + pT - \frac{a}{V}$$

$$\left[\frac{ab}{V^2} \text{ is negligible}\right]$$

$$W = \left[RT + bP_2 - \frac{a}{V_2}\right] - \left[RT + bP_1 - \frac{a}{V_1}\right] - \frac{a}{V_2} + \frac{a}{V_1}$$

$$= b\left[P_2 - P_1\right] + 2a\left[\frac{1}{V_1} - \frac{1}{V_2}\right]$$

But $$V_1 = \frac{RT}{P_1} \text{ and } V_2 = \frac{RT}{P_2}$$

$\therefore$ $$W = b\left[P_2 - P_1\right] + 2a\left[\frac{P_1}{RT} - \frac{P_2}{RT}\right]$$

or $$W = -b\left[P_1 - P_2\right] + \frac{2a}{RT}\left[P_1 - P_2\right]$$

or $$W = (P_1 - P_2)\left(\frac{2a}{RT} - b\right) \qquad ...(i)$$

Suppose the fall in temperature is δT

$$W = JH$$

$$= J[MC_p\delta F]$$

where M is the gram-molecular weight of the gas

$\therefore$ $$JMC_P\ \delta T = \left(P_1 - P_2\right)\left(\frac{2a}{RT} - b\right)$$

or $$\delta T = \left[\frac{P_1 - P_2}{JMC_P}\right] \times \left(\frac{2a}{RT} - b\right) \qquad ...(ii)$$

(i) Since $P_1 - P_2$ is +ve

δT will be +ve if $\left(\frac{2a}{RT} - b\right)$ is +ve

i.e. $$\frac{2a}{RT} > b, \quad \text{or} \quad T < \frac{2a}{Rb} \qquad ...(iii)$$

Therefore, cooling will take place if the temperature of the gas is less than 2a/Rb.

(ii) For δT to be zero, from equation (ii),

$$\frac{2a}{RT} - b = 0$$

or $$T = \frac{2a}{Rb}$$

This temperature is called the temperature of inversion and is represented by T_i

$$T_i = \frac{2a}{Rb} \quad ...(iv)$$

(iii) δT will be negative, if

$$\left(\frac{2a}{RT} - b\right) \text{ is } -ve$$

i.e., $$b > \frac{2a}{RT}$$

or $$T > \frac{2a}{Rb}$$

$$T > T_i.$$

Therefore, heating will take place if the temperature of the gas is more than the temperature of inversion.

Results

(i) If the gas is at the temperature of inversion./then no cooling or heating is observed when it is passed through the porous plug.

(ii) If the gas is at a temperature lower than the temperature of inversion, cooling will take place when it is passed through the porous plug. This is called *regenerative cooling* or *Joule-Kelvin cooling*. This principle has been used in the liquefaction of the so-called permanent gases like nitrogen, oxygen, hydrogen and helium.

(iii) If the gas is at a temperature higher than the temperature of inversion, instead of cooling, heating is observed when the gas is passed through the porous plug.

5.47 RELATION BETWEEN BOYLE TEMPERATURE, TEMPERATURE OF INVERSION AND CRITICAL TEMPERATURE

Temperature of inversion,

$$T_i = \frac{2a}{Rb} \qquad ...(i)$$

Boyle temperature,

$$T_B = \frac{a}{Rb} \qquad ...(ii)$$

Critical temperature,

$$T_e = \frac{8a}{27\ Rb} \qquad ...(iii)$$

From (i) and (ii)

$$T_i = 2T_B \qquad ...(iv)$$

From (i) and (iii)

$$\frac{T_i}{T_e} = \frac{2a}{Rb} \cdot \frac{27\,Rb}{8a}$$

$$= \frac{27}{4} = 6.75$$

The experimental value of T_i/T_e for actual gases is just less than 6.

It means that the temperature of inversion is very much higher than the critical temperature. For hydrogen T_i = 190K and T_e = 33K As $T_i \geq T_e$, the methods employing regenerative cooling (Joule-Kelvin cooling) are preferred to those employing the initial cooling of the gas below the critical temperature.

SOLVED EXAMPLES

Example 1:

A lead bullet strikes a target with a velocity of 480 m/s. If the bullet falls dead, calculate the rise in temperature assuming that all the heat developed is equally shared between, it and the target. (Sp. heat of lead = 0.03)

Solution:

Suppose mass of the bullet = m grams

Rise in temperature = θ

Velocity $v = 480$ m/s

$= 48,000$ cm/s

Work done $= 1/3\ mv^2$

$= 1/2\ m\ (48.000)^2$ ergs

$$H = \frac{W}{J}$$

$$= \frac{m\ (48,000)^2}{2 \times 4.2 \times 10^7} \text{ cals.}$$

Half of this heat energy is used to raise the temperature of the bullet.

$$\therefore \text{ Useful heat } H = \frac{m\ (48,000)^2}{2 \times 2 \times 4.2 \times 10^7} \text{ cals.}$$

$$mS\theta = \frac{m\ (48,000)^2}{2 \times 4.2 \times 10^7}$$

$$\theta = \frac{48,000 \times 48,000}{0.03 \times 4.2 \times 10^7}$$

$$= \mathbf{457.14°C}$$

Example 2:

Calculate the increase in energy per atom of aluminium in ergs when the temperature of a piece of aluminium increases by 1°C (27 g of aluminium contains 6×10^{23} atoms and sp. heat of aluminium = 0.22).

Solution:

Heat required to raise the temperature of 27 grams of aluminium by 1°C

$= 27 \times 0.22 \times 1$ cal

Energy gained by 6×10^{23} atoms of aluminium

$= 27 \times 0.22 \times 4.2 \times 10^7$ ergs

Increase in energy per atom of aluminium

$$= \frac{27 \times 0.22 \times 4.2 \times 10^7}{6 \times 10^{23}} \text{ ergs}$$

$$= \mathbf{4158 \times 10^{-19}\ ergs}$$

Example 3:

What will be the rise in temperature of water if it falls from a height of 50 metres, assuming that all the energy is used in heating the water?

Solution:

We have

h as 50 metres = 5,000 cm

Suppose, rise in temperature = θ°C

Mass of water = m

Work done = W = mgh

$= m \times 980 \times 5{,}000$ ergs

Heat produced = H = mSθ

$= m \times 1 \times \theta$ calories

$$W = JH$$

$$m \times 980 \times 5{,}000 = 4.2 \times 10^7 \times m \times \theta$$

$$\theta = \frac{980 \times 5{,}000}{4.2 \times 10^7}$$

$$= \mathbf{0.117°C.}$$

Example 4:

Calculate the number of molecules in one c.c. of oxygen at N.T.P. using the following data:

Solution:

Density of mercury = 13.6 g/cm^3

R.M.S. velocity of oxygen molecules at 0°C

$= 4.62 \times 10^4$ cm/s

Mass of one molecule of oxygen = 52.8×10^{-24}g

$$P = \frac{1}{3}\rho C^2$$

Let the mass of each molecule be m and the number of molecules in one cc a m

$$\therefore \quad P = \frac{1}{3} mnC^2$$

$$n = \frac{3P}{mC^2}$$

Here $P = 76 \times 13.6 \times 980$ dynes/cm^2

$m = 52.8 \times 10^{-24}$g

$C = 4.62 \times 10^4$ cm/s

$$\therefore \quad n = \frac{3 \times 76 \times 13.6 \times 980}{52.8 \times 10^{-24} \times (4.62 \times 10^4)^2}$$

$$\mathbf{n = 2.697 \times 10^{19}.}$$

Example 5:

At what Celsius temperature will oxygen molecules have the same root mean square velocity as that of hydrogen molecules at –100°C?

Solution:

The energy of a gas molecule is,

$$\frac{1}{2} mC^2 = \frac{3}{2} kT$$

For hydrogen molecules

$$\frac{1}{2} m_1C_1^2 = \frac{3}{2} kT_1 \quad ...(i)$$

For oxygen molecules

$$\frac{1}{2} m_1C_2^2 = \frac{3}{2} kT_2 \quad ...(ii)$$

Dividing (i) by (ii)

$$\frac{m_1C_1^2}{m_2C_2^2} = \frac{T_1}{T_2} \quad ...(iii)$$

Here $C_1 = C_2$

$T_1 = -100°C$

$= -100 + 273$

$= 173$ K

$T_2 = ?$

$$\frac{m_2}{m_1} = 16$$

From equation (iii)

$$T_2 = \frac{m_2 C_2^2 T_1}{m_1 C_1^2}$$

$$T_2 = 16 \times 173$$

$$= 2768 \text{ K}$$

$$T_2 = 2768 - 273$$

$$= \mathbf{2495°C}$$

Example 6:

Calculate the RMS velocity of the oxygen molecules at 27°C.

Solution:

First calculate the RMS velocity of oxygen at N.T.P.

$$C = \sqrt{\frac{3P}{\rho}}$$

Here $P = 76 \times 13.6 \times 980$ dynes/cm^2

$\rho = 16 \times 0.000089$

$$\therefore \quad C = \sqrt{\frac{3 \times 76 \times 13.6 \times 980}{16 \times 0.000089}}$$

$$C = 4.6 \times 10^4 \text{ cm/s}$$

Let the RMS velocity of the molecules at 27°C be C_1

$$\therefore \quad \frac{C_1}{C} = \sqrt{\frac{T_1}{T}}$$

$$C_1 = C \times \sqrt{\frac{T_1}{T}}$$

Here $C = 4.6 \times 10^4$ cm/s

$T = 273$ K

$$T_1 = 27°C$$
$$= 27 + 273$$
$$= 300 \text{ K}$$
$$C_1 = 4.6 \times 10^4 \sqrt{\frac{300}{273}}$$
$$\mathbf{C_1 = 4.84 \times 10^4 \text{ cm/s}}$$

Example 7:

Calculate the volume occupied by 3.2 grams of oxygen at 76 cm of Hg and 21aC.

Solution:

Here $P = 76 \times 13.6 \times 980$ dynes/sq cm

$$T = 27 + 273$$
$$= 306 \text{ K}$$
$$R = 8.31 \times 10^{7'} \text{ ergs/g mol-K.}$$
$$PV = RT$$
$$V = \frac{8.31 \times 10^7 \times 300}{76 \times 13.6 \times 980} \text{ cc per g} - \text{mol}$$
$$V = 24610 \text{ cc per g mol}$$

Volume for 3.2 g of oxygen = v

$$= \frac{24610 \times 3.2}{32}$$
$$\mathbf{= 2461 \text{ cc.}}$$

Example 8:

Show that n, the number of molecules per unit volume of an ideal gas is given by

$$n = \frac{PN}{RT}$$

where N is Avogadrc's number.

Solution:

For an ideal gas, for one gram molecule of a gas,

$$PV = BT$$

But $$R = Nk$$

$$PV = NkT.$$

Let n be the number of molecules per cc. In that case,

$$P = nkT$$

$$\therefore \quad n = \frac{P}{kT}$$

But $$k = \frac{R}{N}$$

$$\therefore \quad n = \frac{PN}{RT}$$

Example 9:

Calculate the number of molecules in one cubic metre of an ideal gas at N.T.P.

Solution:

Let the number of molecules per cc be n.

$$n = \frac{PN}{RT}$$

Number of molecules in one cubic metre volume

$$x = n \times 10^6$$

$$x = \frac{PN \times 10^6}{RT}$$

Here $$P = 76 \times 13.6 \times 980 \text{ dynes/cm}^2$$

$$N = 6.023 \times 10^{23}$$

$$R = 8.31 \times 10^7 \text{ erg/g mol-K}$$

$$T = 273 \text{ K}$$

$$x = \frac{76 \times 13.6 \times 980 \times 6.023 \times 10^{23} \times 10^6}{8.31 \times 10^7 \times 273}$$

$$\mathbf{x = 2.688 \times 10^{25}.}$$

Example 10:

Calculate the number of molecules in one litre of an ideal gas at 136.6°C temperature and 3 atmospheres pressure.

Solution:

Let the number of molecules per cc = n

$$n = \frac{PN}{RT}$$

Number of molecules in one litre,

$$x = n \times 10^3$$

$$x = \frac{PN \times 10^3}{RT}$$

Here $P = 3 \times 76 \times 13.6 \times 980$ dynes/cm^2

$N = 6.023 \times 10^{23}$

$R = 8.31 \times 10^7$ ergs/g mol-K

$T = (273 + 136.5)$

$= 409.5$ K

$$\therefore \quad x = \frac{3 \times 76 \times 13.6 \times 980 \times 6.023 \times 10^{23} \times 10^3}{8.31 \times 10^7 \times 409.5}$$

$$\mathbf{x = 5.376 \times 10^{23}.}$$

Example 11:

The number of molecules per cc of a gas if 2.7×10^{19} at N.T.P. Calculate the number of molecules per cc of the gas.

(i) at 0°C and 10^{-6} mm pressure of mercury and

(ii) at 89°C and 10^{-6} mm pressure mercury.

Solution:

For a unit volume of a gas

$$P = \frac{1}{3} mnC^2$$

(i) At 0°C, R.M.S. velocity is equal to C

At N.T.P.

$$P_1 = \frac{1}{3} m n_1 C^2 \quad \text{...(i)}$$

At 0°C and 10^{-6} mm pressure

$$P_2 = \frac{1}{3} m n_2 C^2 \quad \text{...(ii)}$$

From (i) and (ii)

$$\frac{P_1}{P_2} = \frac{n_1}{n_2}$$

or $$n_2 = \frac{n_1 \times P_2}{P_1}$$

Here $n_1 = 2.7 \times 10^{19}$

$P_1 = 76 \times 13.6 \times 980$ dynes/cm^2

$P_2 = 10^{-6}$ mm of Hg

$= 10^{-7}$ cm of Hg

$P_2 = 10^{-7} \times 13.6 \times 980$ dynes/cm^2

$$n_2 = \frac{2.7 \times 10^{19} \times 10^{-7} \times 13.6 \times 980}{76 \times 13.6 \times 980}$$

$$= 3.553 \times 10^{10}$$

(ii) Let the R.M.S. velocity at 0°C be C_2 and at 39°C be C_3. Number of molecules per cc at 0°C and 10"' mm of Hg pressure

$$= n_2$$

and at 39°C and 10^{-6} mm of Hg pressure

$$= n_3$$

$$P = \frac{1}{3} mn^2$$

Here pressure is the same in both the cases

$$\therefore \quad \frac{1}{3} m n_2 C_2^2 = \frac{1}{3} m n_3 C_3^2$$

$$\therefore \quad n_2 C_2^{\,2} = n_3 C_3^{\,2}$$

$$n_3 = n_2 \frac{C_2^2}{C_3^2}$$

But $C \propto \sqrt{T}$

$$\therefore \quad \frac{C_2^2}{C_3^2} = \frac{T_2}{T_3}$$

$$\therefore \quad n_3 = \frac{n_2 \times T_2}{T_3}$$

Here $n_2 = 3.553 \times 10^{10}$

$T_2 = 0°C$

$= 273$ K

$T_3 = 39°C$

$= 273 + 39$

$= 312$ K

$$\therefore \quad n_3 = \frac{3.553 \times 10^{10} \times 273}{312}$$

$$\mathbf{n_3 = 3.109 \times 10^{10}.}$$

Example 12:

Calculate the total random kinetic energy of one gm-molecule of oxygen at 300 K.

Solution:

Total random kinetic energy per gram-molecule of oxygen

$$= \frac{1}{2} mC^2 \times N$$

$$= \frac{3}{2} kT.N$$

$$= \frac{3}{2} . \frac{R}{N} T.N$$

$$= \frac{3}{2} RT$$

$$= \frac{3}{2} \; 8.3 \times 10^7 \times 300$$

$$= 3.735 \times 10^{10} \text{ ergs}$$

$$= \mathbf{3735 \text{ joules.}}$$

Note: The total random kinetic energy for one gram-molecule of any gas *i.e.*, hydrogen, oxygen, helium, nitrogen, air etc., is the same at the same temperature.

Example 13:

Calculate the average kinetic energy of a molecule of a gas at a temperature of 300 K.

Solution:

Average K.E. of a molecule

$$= \frac{1}{2} mC^2 = \frac{3}{2} kT$$

Here k is Boltzmann's constant.

$$k = 1.38 \times 10^{-16} \text{ erg/molecule-deg}$$

Average K.E. of a molecule = $3/2 \times 1.38$ h $10^{-16} \times 300$

$$= \mathbf{6.21 \times 10^{-14} \text{ ergs}}$$

Note: The average kinetic energy of a molecule of any gas *i.e.*, hydrogen, oxygen, helium, nitrogen, air, etc., is the same at the same temperature.

Example 14:

Calculate the mean translational kinetic energy per molecule of a gas at 727°C, given R = 8.32 joules/mole-K.

Solution:

Avogadro's number

$$N = 6.06 \times 10^{23}$$

$$T = 627 + 273 = 1000 \text{ K}$$

Here $R = 8.32$ joules/mol-K

$$N = 6.06 \times 10^{23}$$

$$k = \frac{R}{N} = \left(\frac{8.32}{6.06 \times 10^{23}}\right) \text{joules/molecule} - K$$

Mean translational kinetic energy per molecule = 3/2 kT

$$= \frac{3 \times 8.32 \times 1000}{2 \times 6.06 \times 10^{23}}$$

$$= \mathbf{2.059 \times 10^{-20}} \textbf{ joule}$$

Example 15:

Calculate the total random kinetic energy of one gram of nitrogen at 300 K.

Solution:

Total random energy for one gram molecule of nitrogen

$$= \frac{3}{2} RT$$

Total random energy for 1 gram of nitrogen

$$= \frac{3RT}{2M}$$

where the molecular weight of nitrogen M = 28 g

$$E = \frac{3RT}{2M}$$

$$= \frac{3 \times 8.3 \times 10^7 \times 300}{2 \times 28}$$

$$= 133.4 \times 10^7 \text{ ergs}$$

$$= \mathbf{133.4 \text{ joule.}}$$

Note: The total random kinetic energy for 1 gram of a gas is different for different gases at the same temperature.

Example 16:

Calculate the total random kinetic energy of 2 K of helium at 200 K.

Solution:

Energy for 1 g of helium

$$= \frac{3RT}{2M}$$

Energy for 2 g of helium

$$= \frac{2 \times 3 \times RT}{2M}$$

$$= \frac{3RT}{M}$$

$$= \frac{3 \quad 8.3 \times 10^7 \times 200}{4}$$

$$= 1245 \times 10^7 \text{ ergs}$$

$$= \mathbf{1245 \text{ joules.}}$$

Example 17:

Calculate the root mean square velocity of a molecule of mercury vapour at 300 K.

Solution:

Mean kinetic energy of One molecule of-mercury

$$= \frac{1}{2} mC^2 = \frac{3}{2} kT$$

Let N be the Avogadro's number.

Then $\frac{1}{2} mNC^2 = \frac{3}{2} kNT$

$$\frac{1}{2} MC^2 = \frac{3}{2} RT$$

$$C = \sqrt{\frac{3RT}{M}}$$

Here the molecular weight of mercury, M = 221

$$\therefore \qquad C = \sqrt{\frac{3 \; 8.3 \times 10^7 \times 300}{221}}$$

$$= \mathbf{1.93 \times 10^4 \text{ cm/s.}}$$

Example 18:

With what speed would one gram molecule of oxygen at 300 K be moving in order that the translational kinetic energy of its centre of mass is equal to the total random kinetic energy of all its molecules? Molecular weight of oxygen (M) = 32.

Solution:

Total random kinetic energy of 1 gram-molecule of oxygen

$$= \frac{3}{2}RT = \frac{3}{2} \times 8.3 \times 10^7 \times 300$$

$$= 3.735 \times 10^{10} \text{ ergs}$$

Kinetic energy of M grams of oxygen moving with a velocity v

$$= \frac{1}{2}Mv^2$$

$$\therefore \quad = \frac{1}{2}Mv^2 = 3.735 \times 10^7$$

$$\frac{1}{2} \times 32 \times v^2 = 3.735 \times 10^{10}$$

$$v^2 = \frac{3.735 \times 10^{10}}{16}$$

$$\mathbf{x = 4.8 \times 10^4 \text{ cm/s.}}$$

Example 19:

Calculate the temperature at which the r.m.s. velocity of a hydrogen molecule will be equal to the speed of the earth's first satellite (i.e. v as 8 km/s).

Solution:

Energy for 1 gram molecule of hydrogen

$$= \frac{1}{2}Mv^2 = \frac{3}{2}RT$$

$$T = \frac{Mv^2}{3R}$$

$$= \frac{2 \times (8 \times 10^5)^2}{3 \times 8.3 \times 10^7}$$

$$\mathbf{= 5.14 \times 10^3 \text{ K.}}$$

Example 20:

At what temperature, pressure remaining constant, will the r.m.s. velocity of a gas be half its value at 0°C?

Solution:

We have $\frac{C_2}{C_1} = \sqrt{\frac{T_2}{T_1}}$

Here $C_2 = \frac{C_1}{2}$

$T_2 = ?$

$T_1 = 273\ K$

$\therefore \quad \frac{1}{2} = \sqrt{\frac{T_2}{273}}$

$\frac{1}{4} = \frac{T_2}{273}$

$T_2 = \frac{273}{4} = 68.25\ K$

or $\mathbf{T_2 = -204.75°C.}$

Example 21:

In an experiment, the viscosity of the gas was found fobs 1.66 × 10^{-4} dynes/cm² per unit velocity gradient. The R.M.S. velocity of the molecules is 4.5 × 10^4 cm/s. The density of the gas is 1.25 grams per litre.

Calculate (i) the mean free path of the molecules of the gas, (ii) frequency of collision and (iii) molecular diameter of the gas moleeules.

Solution:

Here $\eta = 1.66 \times 10^{-4}$ units

$C = 4.5 \times 10^4$ cm/s

$\rho = 1.25 \times 10^{-3}$ g/cc

(i) $\lambda = \frac{3\eta}{\rho C}$

$$\lambda = \frac{3 \times 1.66 \times 10^{-4}}{1.25 \times 10^{-3} \times 4.5 \times 10^{4}}$$

$$\lambda = \mathbf{9. \times 10^{-6}\ cm.}$$

(ii) Frequency of collision = number of collisions per second

$$= \frac{\text{R.M.S. velocity}}{\text{Mean free path}}$$

$$\therefore \quad N = \frac{C}{\lambda}$$

$$N = \frac{4.5 \times 10^{4}}{9 \times 10^{-6}}$$

$$\mathbf{N = 5 \times 10^{9}}$$

(iii) Avogadro's number

$$= 6.023 \times 10^{23}$$

Number of molecules per cc = n

$$= \frac{6.028 \times 10^{23}}{22400}$$

According to Maxwell's relation

$$\lambda = \frac{1}{\sqrt{2}\ .\ \pi d^2 n}$$

$$\therefore \quad d = \frac{1}{\sqrt{1.414 \times \pi \times n \times \lambda}}$$

$$d = \frac{1}{\sqrt{1.414 \times 3.142 \times \left(\frac{7.023\ 0 + 10^{23}}{22400}\right) \times 9 \times 10^{-6}}}$$

$$\mathbf{d = 3 \times 10^{-6}\ cm.}$$

Example 22:

In an experiment the viscosity of the gas was found to be 2.25 × 10^{-4} CGS unite. The RMS velocity of the molecules is 4.5 × 10^{4} cm/s. The density ofthe gas is 1 gram per litre. Calculate the mean free path of the molecules.

Solution:

Here $\eta = 2.25 \times 10^{-4}$ CGS units

$C = 4.5 \times 10^{4}$ cm/s

$\rho = 1$ g/litre

$\therefore \quad \rho = 10^{-2}$ g/cc

$$\lambda = \frac{3\eta}{\rho C}$$

$$\lambda = \frac{3 \times 2.25 \times 10^{-4}}{10^{-3} \times 4.5 \times 10^{4}}$$

$$\lambda = \mathbf{15 \times 10^{-6}\ cm.}$$

Example 23:

Calculate the mean free path of a gas molecule, given that the molecular diameter is 2×10^{-8} cm and the number of molecule per cc is 3×10^{19}.

Solution:

We have $\lambda = \dfrac{1}{\pi d^2 n}$

$$= \frac{1}{3.14\ 0 + (2 \times 10^{-8})^2\ 0 + 3 \times 10^{19}}$$

$$\lambda = \mathbf{3 \times 10^{-65}\ cm.}$$

Note: The mean free path is less than the wavelength of light in the visible spectrum.

Example 24:

Calculate the mean free path of gas molecules in a chamber of 10^{-6} mm of mercury pressure, assuming the molecular diawetor to be 2Å One gram molecule of the gas occupies 22.4 litres at N.T.P. Take the temperature of the chamber to be 273 K.

Solution:

At 760 mm Hg pressure and 273 K. temperature, the number of molecules in 22.4 litres of a gas

$$= 6.023 \times 10^{23}$$

Therefore, the number of molecules per cm^3 in the chamber at 10^{-6} mm pressure and 273 K temperature

$$n = \frac{6.023 \times 10^{23} \times 10^{-6}}{22400 \times 760}$$

$$n = 3.538 \times 10^{10} \text{ molecules/cm}^2$$

$$d = 2Å = 2 \times 10^{-6} \text{ cm}$$

Mean free path,

$$\lambda = \frac{1}{\pi d^2 n}$$

$$= \frac{1}{3.14 \times (2 \times 10^{-8})^2 \times 3.538 \times 10^{10}} = 2.25 \times 10^4 \text{ cm}$$

Example 25:

Calculate the Vander Waals constants for dry air, given that

T_e = *132 K*, P_e = 37.2 atmospheres,

R per mole = 82.07 cm³ atmos K⁻¹.

Solution:

Here P_e = 37.2 atomospheres

T_e = 132 K

R = 82.07 cm^3 atoms K^{-1}

(i) $$a = \frac{27}{64} \frac{R^2 T_e^2}{P_e}$$

$$a = \left(\frac{27}{64}\right) \frac{(82.07)^2 (132)^2}{37.2}$$

or **a = 13.31 × 10⁶ atmos cm⁶.**

EXERCISES

1. What do you understand by degrees of freedom of a molecule in a thermal system? State the law of equipartition of energy. Apply this law to obtain specific heats, C_p and C_v of gases. Compare the values of γ the ratio of C_p and C_v, so obtained with the experimental results in the case of monoatomic, diatomic

and polyatomic gases. Discuss the discrepancies, wherever they occur.

2. Describe the aim of the porous-plug experiment and its predictions. Outline the important results of this experiment.
3. Derive an expression for the change in the temperature of a gas undergoing Joule-Thomson expansion. Discuss the role of inversion temperature in it.
4. Define viscosity of a fluid. Obtain an expression for the viscosity of an ideal gas on the basis of kinetic theory.
5. Calculate the total random kinetic energy of 1 gram molecule of nitrogen at 300 K.
6. Calculate the average kinetic energy of a hydrogen molecule at 27°C.
7. Calculate the total random kinetic energy, of 2 grams of nitrogen at 27°C.
8. Calculate the total random kinetic energy of 8 grams of helium at 200 K.
9. Calculate the r.m.s. velocity of a mercury atom at 1,200 K.
10. With what speed would one gram molecule of hydrogen at 27°C be moving in order that the translational kinetic energy of its centre of mass is equal to the total random kinetic energy of all its molecules? (Molecular weight of hydrogen = 2).
11. Discuss briefly the considerations which led Van der Waals to modify the gas equation. What are the critical constants of a gas? Calculate the values of these constants in terms of the constants of the Van der Waals equation.
12. Show that the pressure exerted by a perfect gas is 2/3 of the Kinetic energy of the gas molecules in a unit volume.
13. Derive and discuss the Van der Waals equation of state of a gas. Mention its defects.
14. Define mechanical equivalent of heat and give its units in the C.G.S. system. Describe fully Searle's friction-cone method of determining the value of J. Discuss the necessary formula and draw a neat diagram of the apparatus.
15. Deduce Van der Waals equation of state. How far does it conform to Andrews' experimental results on carbon dioxide? What is

the importance of Andrews' experiments in the problem of liquefaction of gases?

16. Describe the porous plug experiment. What conclusions have been drawn from it?
17. Describe Joule-Thomson effect and give its theory.
18. Derive Maxwell's distribution law of velocities for gas molecules and discuss its experimental verification.
19. What is meant by Joule-Kelvin effect? How is it experimentally established? How will you interpret the effect?
20. Describe Callendar and Barnes' continuous flow method for finding J.
21. Describe Jaegar and Steinwehr's method for finding the value of J.
22. Describe Holborn's experiments to study the behaviour of gases at high pressure.
23. Describe experiments to find the value of the critical constants of a gas.
24. Derive the relations between the Boyle temperature, temperature of inversion and critical temperature.
25. Derive an expression for the pressure exerted by gas on the basis of kinetic theory.
26. Describe the porous plug experiment and derive an expression for the Joule-Thomson cooling of a gas. What is inversion temperature?
27. Derive an expression for the mean free path of a molecule of a gas and describe a method for its determination.
28. Obtain an expression for the pressure of a gas on the basis of the kinetic theory of gases. Show that, according to the law of equipartition of energy the ratio of the two specific heats of ozone is 1.33.
29. Distinguish between a perfect gas and a real gas. Derive Van der Waals equation of state and use it to obtain the expressions for the critical constants in terms of the constants of the Van der Waals equation.
30. What is the kinetic model of a gas?. How is pressure of a gas explained on this model? Deduce an expression for the pressure?

31. Describe briefly the porous plug experiment. Deduce a theoretical expression for the Joule-Thomson cooling for 1 gram molecule of a gas obeying Van der Waals equation of state. Why does hydrogen show a negative Joule-Thomson effect?

32. Distinguish between an adiabatic process and Joule-Thomson effect. Prove that for a Van der Waals gas the Joule-Thomson coefficient is given by

$$\left(\frac{\delta T}{\delta P}\right)=\frac{1}{C_p}\left[\frac{2a}{RT}-b\right]$$

33. What is Joule-Thomson effect? Derive an expression, for Joule-Thomson cooling. What is temperature of inversion? Prove that it is equal to 2a/Rb.

34. Show from the kinetic theory of gases that the mean kinetic energy of translation of a molecule of a gas is 3/2 kT, where k is the Boltzmann's constant.

35. How is the ideal gas equation modified when mutual attraction between molecules and the finite size of the molecules are taken into consideration? Obtain relations between critical constants and the constants appearing in the Van der Walls equation.

36. Derive the reduced equation of state for a gas starting from Van der Waals equation of state. Show that if the two gases have the same reduced pressure and volume, they also have the same reduced temperature.

37. Starting from elementary ideas set up Van der Waals' equation of state and show how the constants of the equation can be expressed in terms of the critical constants. What is the experimental bearing of this equation?

38. Assuming the various postulates underlying the kinetic theory of gases, show that the mean kinetic energy of translation of a molecule is 3/2 kT.

39. Derive Van der Waals equation for gases and calculate the theoretical values of the critical constants.

40. Show that the pressure exerted by a perfect gas is 2/3 of the mean kinetic energy per unit volume.

41. On the basis of kinetic theory of gases, derive expressions for the thermal conductivity and viscosity of the gas. Hence obtain the relation between the two.

42. Explain the basic principles of the kinetic theory of gases and show that the pressure of an ideal gas is proportional to its density.

43. Deduce Van der Waals equation of state and obtain expressions for critical constants in terms of the constants of the equation.

44. Describe Andrews' experiments on carbon dioxide. Discuss the results obtained. What is the importance of these results in the liquefaction of gases?

45. How do you interpret:

    (i) pressure, and

    (ii) temperature on the basis of the kinetic theory of gases.

46. What is the meaning of mean free path of the molecules of a gas? Show that it is equal to $1/\eta\pi\sigma^2$

47. Give a discussion of Maxwellian distribution of speed, C amongst N molecules per cm^3 enclosed in a chamber at a temperature T.

48. Express Maxwell's law of distribution of speeds in terms of the kinetic energy of the molecules. Hence find the most probable and the average energy of the molecules.

49. Derive an expression for the viscosity of a gas on the basis of the kinetic theory and discuss its dependence on temperature and pressure.

50. Obtain an expression for the thermal conductivity of an ideal gas on the basis of kinetic theory. Discuss its dependence on pressure and temperature.

51. Calculate the temperature at which the r.m.s. velocity of a helium molecule will be equal to the speed of the earth's first satellite *i.e.*, v = 8 km/s.

52. Calculate the mean kinetic energy of a molecule of a gas at 1,000°C. Given,

    $$R = 8.31 \times 10^7 \text{ ergs/gram mol-K}$$

    $$N = 6.02 \times 10^{23}.$$

53. If the density of nitrogen is 1.25 g/litre at N.T.P., calculate the R.M.S. velocity of its molecules.

54. At what temperature is the R.M.S. speed of oxygen molecules twice their R.M.S., speed at 27°C?

55. Calculate the R.M.S. velocity of the molecules of hydrogen at 0°C. Molecular weight of hydrogen = 2.016 and

$$R = 8.31 \times 10^7 \text{ ergs/gram mole °C.}$$

56. Calculate the R.M.S. velocity of the hydrogen molecules at room temperature, given that one litre of the gas at room temperature and normal pressure weighs 0.086 g.

57. Write short notes on:

    (i) Mean free path
    (ii) Joule-Thomson Effect
    (iii) Continuity of state
    (iv) Rowland's experiment for finding J
    (v) Van der Waals equation of state
    (vi) Pressure exerted by an ideal gas
    (vii) Critical constants
    (viii) Degrees of freedom
    (ix) Atomicity of gases
    (x) Maxwell's law of distribution of velocity.
    (xi) Andrews' experiments
    (xii) Amagat's experiments
    (xiii) Halborn's experiments
    (xiv) Behaviour of gases at high pressure
    (xv) Critical point
    (xvi) Corresponding states
    (xvii) Intermolecular attraction
    (xviii) Temperature of inversion
    (xix) Reduced equation of state for a gas
    (xx) Porous plug experiment.

58. Describe Andrews' experiments on carbon dioxide. Discuss the results obtained by him. Show that the liquid and the gaseous states are two distant stages of a continuous phenomenon.

59. Derive on the basis of kinetic theory of gases the laws for an ideal gas.

60. Give the theory of the porous plug experiment.

61. What are the critical constants of a gas? State and explain Van der Waals equation. Calculate the critical constants of gas in terms of the constants of this equation.

62. State the law of equipartition of energy. Prove the law for a perfect gas, whose molecules have n degrees of freedom. Show that for a mono-atomic gas $\gamma = 1.66$ and for a diatomic gas $\gamma = 1.40$.

63. What is Joule-Thomson effect? Obtain an expression for the cooling produced in a Van der Waals gas. Hence explain why hydrogen and helium show heating effect at ordinary temperatures.

64. Explain what you mean by degrees of freedom. State the law of equipartition of energy. Prove that for a perfect gas whose molecules have n degrees of freedom

$$\frac{C_P}{C_V} = 1 + \frac{2}{n}$$

Hence show that for a mono-atomic gas $\gamma = 1.67$, for a diatomic gas $\gamma = 1.4$ and for triatomic gas $\gamma = 1.33$.

65. Derive Van der Waals equation. Deduce expressions for the critical constants in terms of a and b.

66. Deduce from the kinetic theory of gases, an expression for the pressure of a gas. Also prove that

$$PV = RT$$

67. Explain the corrections introduced by Van der Waals in the gas equation. Show that for a gas obeying Van der Waals equation

$$\frac{RT_e}{P_eV_e} = \frac{8}{3}.$$

6

Liquefaction of Gases

6.1 INTRODUCTION

Initially it was thought that air remains in the gaseous state in all temperatures. Celsius and Reaumer used in their respective thermometers. The melting point of ice at N.T.P. as zero. They were of the opinion that nothing can be colder than ice. Later, Fahrenheit found that a mixture of snow and NH_4Cl produced a temperature –18°C. He therefore used this temperature as the zero of his thermometer. Thereafter other freezing mixtures were discovered.

Andrew's experiment can CO_2 showed that below the critical temperature, a gas can be liquefied by mere application of pressure however large the applied pressure may be below the critical temperature the gas is termed as vapour and above the critical temperature it is called a gas.

In 1877, Pictet liquefied oxygen by Cascade process. Linde was able to liquefy air in 1896 by Joule–Kelvin effect. Using the principle of Joule–Kelvin effect, hydrogen and helium were also liquefied.

Gas	Critical Temp.	Boiling Point
Helium	–267.8°C	–269.8°C
Hydrogen	–240°C	–252.8°C
Nitrogen	–146°C	–195.8°C
Oxygen	–118.8°C	–183°C
CO_2	–31.1°C	–78.6°C

6.2 CASCADE PROCESS—LIQUEFACTION OF OXYGEN

In 1878 Pictet was able to liquefy oxygen. Later on K. Onnes modified the apparatus as shown in Fig. 6.1.

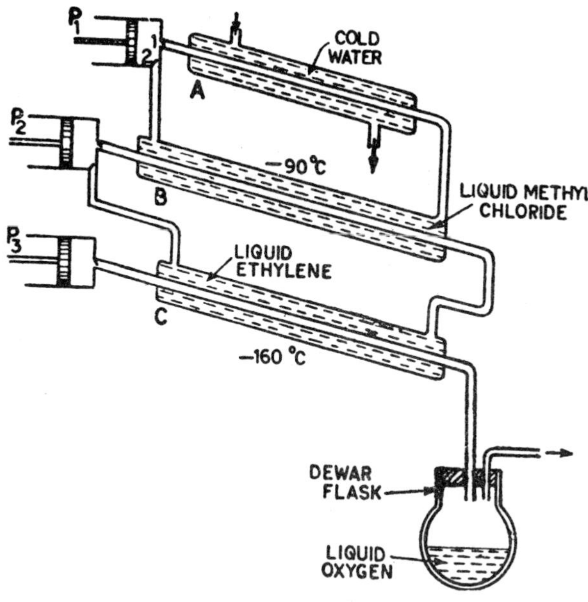

Fig. 6.1

The apparatus consists of compression pumps P_1, P_2, and P_3. A, B and C are the outer jackets containing cold water, liquid methyl chloride and liquid ethylene. The pump P_1 compresses methyl chloride gas and it is cooled by cold water circulating in the jacket A. As the critical temperature of methyl chloride is 143°C, it is liquefied. Liquid methyl chloride circulates in the outer jacket B.

The pump P_2 compresses ethylene gas and it is Surrounded by the outer jacket B containing liquid methyl chloride at about –24°C.

Liquid methyl chloride is allowed to boil under reduced pressure with the help of the pump P_1. The temperature of B reaches –90°C. The critical temperature of ethylene is 10°C. It is cooled to –90°C and due to compression it gets liquefied.

The liquid ethylene in the jacket C is allowed to boil under reduced pressure with the help of the pump P_2 and finally a temperature of

$-160°C$ is reached. The critical temperature for oxygen is $-118.8°C$. Oxygen is compressed to a pressure of about 25 atmospheres and is passed through a spiral tube surrounded by the jacket C. Oxygen gets liquefied and is collected in a Dewar flask. Oxygen in the form of gas in the Dewar flask is circulated back to the pump P_3 and the process is repeated.

Liquefaction of Nitrogen : Liquid oxygen has a normal boiling point $-183°C$. It is allowed to boil under reduced pressure and a temperature of $-218°C$ is reached. The critical temperature of nitrogen is $-146°C$. By adding a fourth unit containing liquid oxygen, nitrogen can be liquefied. Cascade process cannot be used to liquefy hydrogen ($T_C = -240°C$), neon ($T_C = -229°C$) and helium $-268°C$); because in this process the lowest temperature obtainable is $-240°C$.

6.3 LIQUEFACTION OF AIR—LINDE'S PROCESS

In this process, Joule–Kelvin effect is applied. Linde was able to liquefy air in 1896 using this effect. The apparatus used is shown in Fig. 6.2.

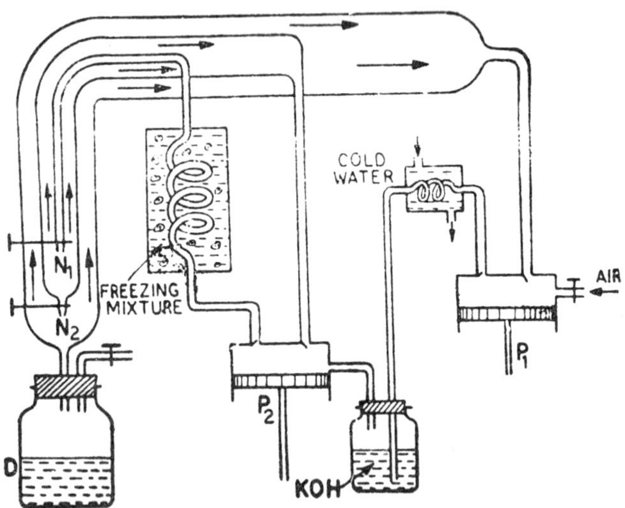

Fig. 6.2

The pump P_1 compresses air to a pressure of about 25 atmospheres and is passed through a tube surrounded by a jacket through which cold water is circulated. This compressed air is passed through KOH solution to remove CO_2 and water vapour.

This air, free from CO_2 and water vapour, is compressed to a pressure of 200 atmospheres by the pump P_2. This air passes through a spiral tube surrounded by a jacket containing a freezing mixture. This cooled air at high pressure and at a temperature of –20°C is allowed to come out of the nozzle. N_1. Joule–Kelvin effect takes place and the incoming air is cooled. The cooled air is circulated to the pump P_2 and is compressed. It passes through the nozzle N_1 and is further cooled. This cooled air is allowed to pass through the nozzle N_2 (from high pressure to low pressure side) and is further cooled. As the process continues, after a few cycles, air gets cooled to a sufficiently low temperature and after coming out of the nozzle N_2 gets liquefied and is collected in the Dewar flask. The unliquefied air is again circulated back to the pump P_1 and the process is repeated.

The whole of the apparatus is packed in cotton wool to avoid any conduction or radiation.

6.4 LIQUEFACTION OF HYDROGEN

Hydrogen cannot be liquefied by Cascade process because it's critical temperature is –240°C. Linde's ordinary apparatus used for the liquefaction of air cannot be used for hydrogen because the temperature of inversion for hydrogen is –83°C. The gas must initially be cooled to a temperature lower than the temperature of inversion or the cooling to take place due to Joule–Kelvin effect. The original apparatus designed by Dewar (1898) was improved later by Travers, Olszewski, Nernst and others (Fig. 6.3). To have complete insulation the whole apparatus is enclosed in an outer Dewar flask L.

Hydrogen under a pressure of 200 atmospheres is passed through a coil immersed in solid CO_2 and alcohol. It enters the coil in the chamber A Where it is further cooled by the outgoing hydrogen. The chamber B contains liquid air and cools hydrogen in the coil E. In the chamber C liquid air is allowed to boil under reduced pressure (10cm of Hg) and hydrogen in the coil F is cooled to a temperature of –200°C. This cooled hydrogen passed through the regenerative coil C and the nozzle N. Hydrogen is cooled further due to Joule–kelvin effect. The cooled hydrogen coming from the nozzle N is allowed to circulate back to the pump as shown in Fig. 6.3. The process of regenerative cooling continues and after some time hydrogen gets liquefied and is collected in the Dewar flask D.

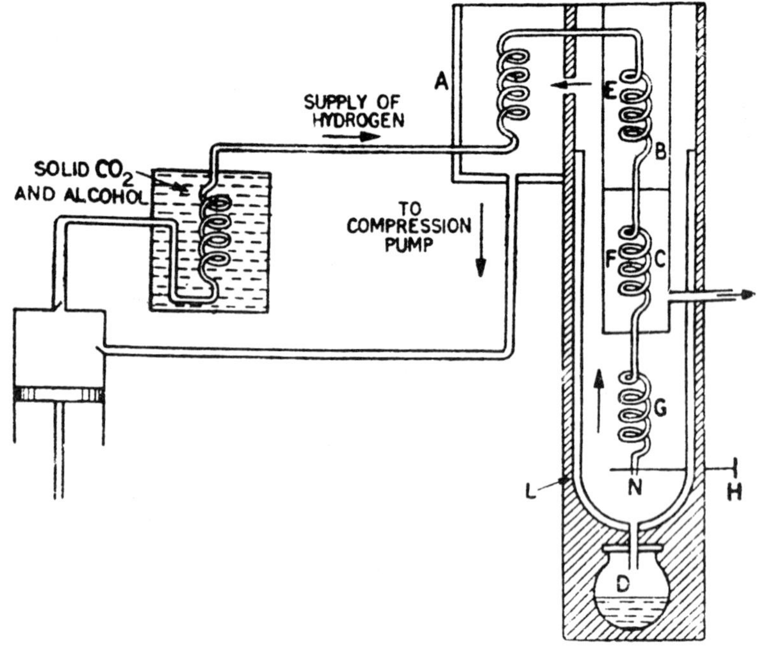

Fig. 6.3

6.5 SOLIDIFICATION OF HYDROGEN

By boiling liquid hydrogen under reduced pressure (10 mm of Hg) Dewar was able to solidify hydrogen. Liquid hydrogen was contained in a double walled thermos tube which was immersed in an outer bath containing liquid hydrogen. The pressure inside the tube was reduced and the temperature of hydrogen in the tube decreased below –259°C. Liquid hydrogen did not solidify but was only super-cooled below the freezing point. By allowing a little trace of air to leak into the apparatus it was possible to obtain solid hydrogen. Solid hydrogen is white in colour.

6.6 CLAUDE'S PROCESS—LIQUEFACTION OF AIR

The experimental arrangement is shown in Fig. 6.4. Air, free from CO_2 and water vapour, is compressed to a pressure of 40 atmospheres by the compressor P_1. Initially air is cooled by passing it through a coil immersed in a freezing mixture. At the point A, 80% of the compressed air goes to the expansion chamber and 20% to the heat exchanger. In the expansion chamber air expands, does external work and drives the piston P_2 outwards. Due to adiabatic expansion, air gets cooled. The expansion chamber and the compressor are coupled so that when P_2 moves outwards, P_1 moves inwards and compresses the air in the compressor.

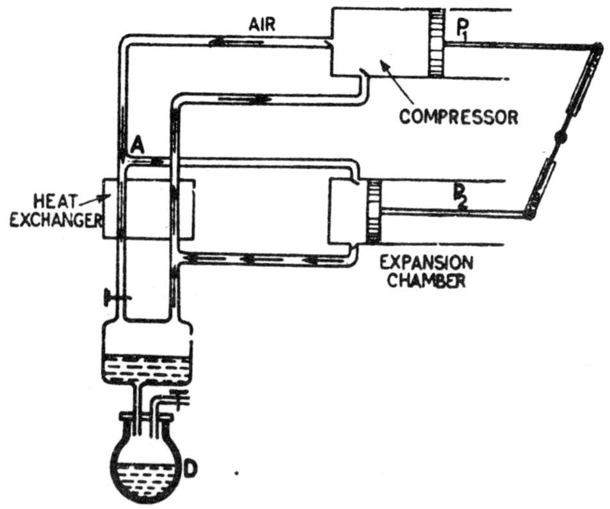

Fig. 6.4

The cooled air from the expansion chamber passes through the heat exchanger and takes heat from it. 20% of air, passing through the tube in the heat exchanger, gives heat and consequently gets cooled. This regenerative cooling process continues and when the temperature of air reaches the liquefaction temperature, it gets liquefied and is collected in the Dewar flask D.

The main problem in this process is that of solidification of lubricants at low temperatures. Petroleum ether is found to be a useful lubricant at low temperatures. Once air gets liquefied, it itself serves as a lubricant.

This process is complicated apart from the problem of lubricants. Linde process is commercially used for the liquefaction of air.

6.7 LIQUEFACTION OF HELIUM—K ONNES METHOD

K. Onnes liquefied helium in 1908. The temperature of inversion of helium is –240°C. Kapitza has liquefied helium by precooling it and passing it through a coil surrounded by a bath containing liquid hydrogen boiling under reduced pressure. In this way helium was cooled to –258°C.

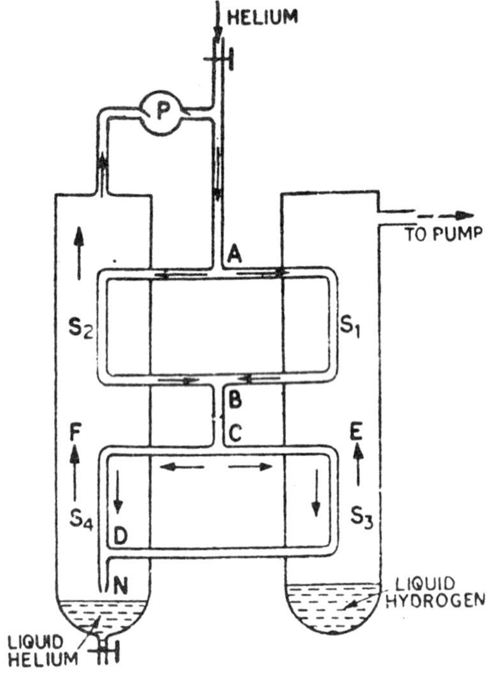

Fig. 6.5

The apparatus used for liquefaction of helium is shown in Fig. 6.5. Helium gas at a pressure of 40 atmospheres enters the spiral tube at A. It is divided into two portions. Helium passing through the spiral in E is cooled because it is "surrounded by hydrogen boiling under reduced pressure. The other portion of helium gas passing through the spiral in F is cooled due to outgoing fooled helium gas. Similar processes take place in the spirals S_3 and S_4. The process is repeated and when the temperature of helium is sufficiently low, it gets liquefied after passing through the nozzle N. At N cooling takes place due to Joule–Kelvin effect. This outgoing helium is compressed again by P and fed back to the spirals S_1 and S_2. Liquefied helium is collected in the Dewar flask. The whole apparatus is surrounded by Dewar flasks to provide perfect heat insulation.

Simon has been able to liquefy helium by using activated charcoal. When helium is adsorbed by charcoal, heat is evolved and when it is desorbed (removed), the temperature of the gas falls. Helium is adsorbed in large quantities by activated charcoal immersed in liquid hydrogen. The apparatus is enclosed in an evacuating vessel to minimize heat exchanges with the surroundings. On pumping off helium, its temperature falls considerably and it gets liquefied.

6.8 HELIUM I AND HELIUM II

K. Onnes observed no sign of solidification when liquid helium was cooled at ordinary pressures. It was found that at a temperature of 2.19 K, the liquid which was contracting when cooled, began to expand (Fig. 6.6).

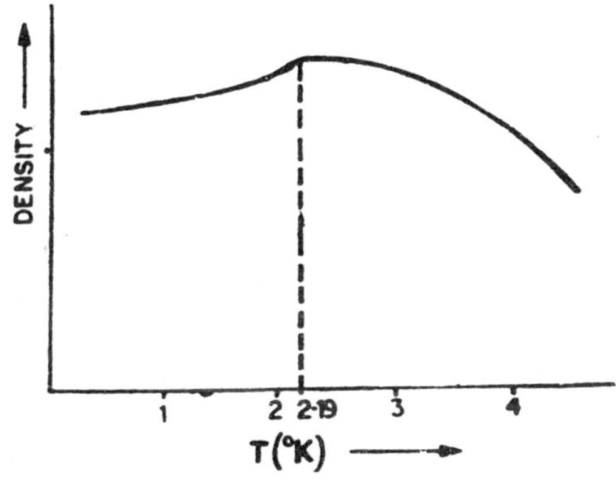

Fig. 6.6

The specific heat of liquid helium increases up to 2.19 K and at this temperature there is a sudden and abnormal increase in the specific heat. Beyond 2.19 K, the specific heat first decreases and then increases (Fig. 6.7).

The specific heat–temperature graph resembles λ and hence this temperature at which the specific heat changes abruptly (2.19 K) is called the λ–point. Liquid helium above 2.19 K is called helium I because it behaves in a-normal way and below 2.19 K it is called helium II because of its abnormal properties.

Viscosity of liquid helium I decreases with decrease in temperature and this property is contrary to the property of a liquid but it resembles that of a gas. The viscosity of helium II is practically zero and it can flow rapidly through narrow capillary tubes.

The thermal conductivity of helium II has an abnormally high value and it is many times more than that of copper and silver.

Helium II forms a thin film on all solid surfaces This film is called the *Roiling film.* It is through this film that helium II flows from one vessel to another. This behaviour of helium II is very peculiar. It can creep into a vessel when the vessel is lowered into the liquid [Fig. 6.8 (i)]. On the other hand when the vessel containing helium II is taken out of the liquid, it creeps out of the vessel and continues to flow until the level outside and inside is the same. [Fig. 6.8 (ii)]. Even when the vessel is completely out of the liquid, helium II creeps out and flows down the outer surface of the vessel [Fig. 6.8 (iii). It continues to flow until the vessel is empty.

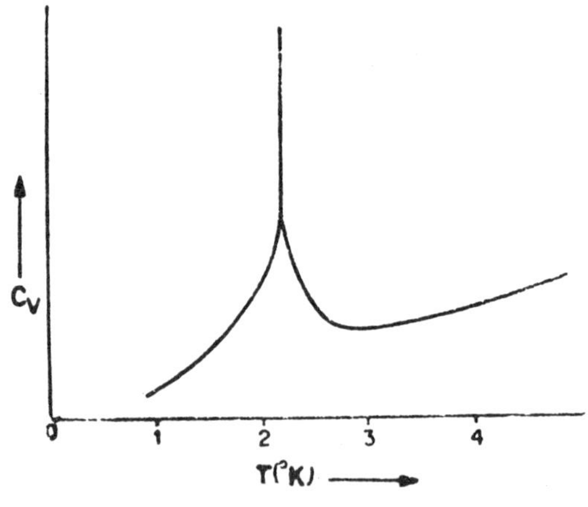

Fig. 6.7

Helium II has a higher heat of vaporization and smaller surface tension. The large specific heat anomaly of liquid helium at 2.19 K is due to rapid decrease of its entropy with decreasing temperatures. This is a complicated phenomenon and statistical mechanics cannot be applied. However, Bose-Einstein statistics can be applied in this case.

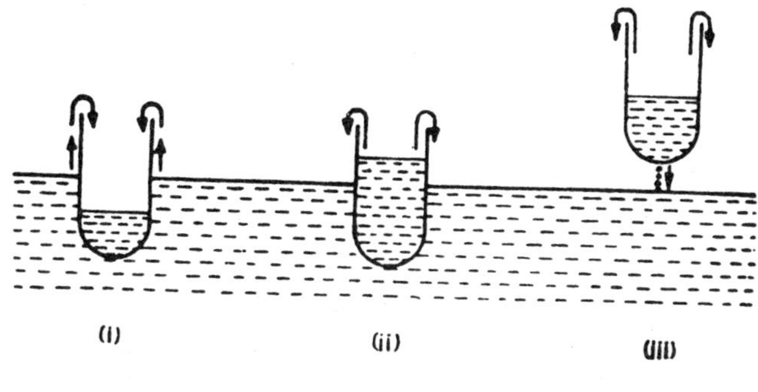

Fig 6.8

6.9 PRODUCTION OF LOW TEMPERATURES

Quite long ago is was noticed that there is a lower limit for a temperature scale. There is no limit for high temperatures but there is a limit for low temperature. The lowest temperature corresponds to 1K. (–273.16°C) called the absolute zero temperature. Attempts have been made to reach absolute zero temperature.

Temperatures below zero degree centigrade can be obtained with the help of freezing mixtures. Temperatures up to –65°C can be obtained with KOH and ice. With the liquefaction of gases temperatures lower than –65°C could be achieved. With liquid helium boiling under normal pressure a temperature of –268.9°C can be reached. By boiling liquid helium under reduced pressure temperature of the order of 1K could be obtained. With liquid helium (isotope, He^3) boiling under reduced pressure a temperature of 0.4 K can be reached. Temperatures below 0.4 K can be reached by 'adiabatic demagnetisation methods' due to Debye (1926) and Giauque (1927).

6.10 ADIABATIC DEMAGNETISATION

Debye, Giauque and Macdougall were able to produce temperatures below 1 K with the help of gadolinium sulphate. Haas and Kramers used the magnetic balance for a number of paramagnetic substances. They found that potassium and chromium alum give a much lower temperature.

The apparatus used is shown in Fig. 6.9. The paramagnetic salt is suspended in a vessel, which is surrounded by liquid helium. Liquid helium is boiled under reduced pressure., It is surrounded be a Dewar flask containing liquid hydrogen. The salt is in contact with the helium gas. A magnetic field of the order of 30,000 gauss is applied.

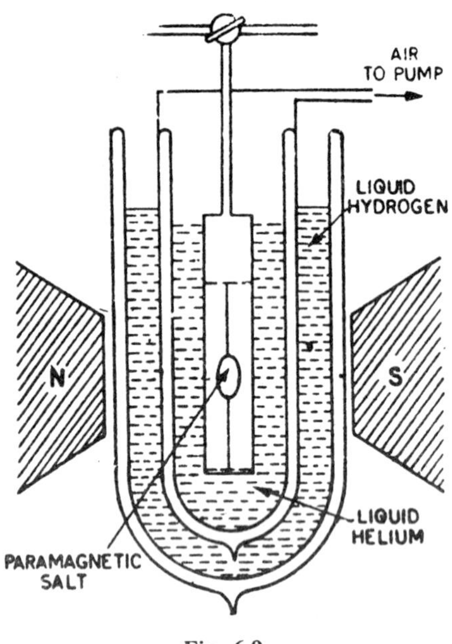

Fig. 6.9

When the magnetic field is switched on, the temperature of the salt rises. But the heat is conducted by the helium gas rapidly and the temperature of salt falls to the original temperature off the helium bath. The helium gas is pumped out and the salt is thermally isolated.

Now the magnetic field is switched off and the temperature of the salt falls due to adiabatic demagnetisation.

To measure the temperature of the salt, the susceptibility is measured with the help of magnetic thermometers. According to Curie's law.

χT is constant

$$\therefore \quad \frac{\chi_1}{\chi_2} = \frac{T_2}{T_1}$$

χ_1 is the sesceptibility at temperature T_1 of the helium bath. χ_2 is the susceptibility after adiabatic demagnetisation at temperature T_2.

$$\therefore \quad T_2 = \frac{\chi_1}{\chi_2} T_1$$

The thermodynamical behaviour of a paramagnetic crystal, placed in a magnetic field, depends also on the strength of the magnetising field in addition to pressure and volume. The usual relation

$$\delta H = dU + PdV$$

is modified in the form

$$\delta H = dU + PdV - B.dI \qquad ...(i)$$

Here B is the magnetic flux density and dI is the change in the intensity of magnetisation per gram molecule. Therefore

$$I = \chi VB \qquad ...(ii)$$

where χ is the susceptibility per unit volume

From the second law of thermodynamics, the change in entropy is given by

$$TdS = dU + PdV - BdI$$

or
$$dU = TdS + BdI - PdV \qquad ...(iii)$$

If x and y are a pair of independent thermodynamical variables, then from the condition

$$\frac{\partial^2 U}{\partial x \partial y} = \frac{\partial^2 U}{\partial y \partial x}, \text{ we get}$$

$$\frac{\partial(T,S)}{\partial(x,y)} + \frac{\partial(B,I)}{\partial(x,y)} = \frac{\partial(P,V)}{\partial(x,y)} \qquad ...(iv)$$

In the adiabatic demagnetisation experiments, the change in volume is negligible. It means $\delta V = 0$ *i.e.* V = constant.

$$\therefore \quad \frac{\partial(T,S)}{\partial(x,y)} + -\frac{\partial(B,I)}{\partial(x,y)} \qquad ...(v)$$

This is a general equation connecting T, S, B and I.. The various thermodynamical relations can be derived by choosing any two out of these four variables T, S, B and I.

(1) Taking x = B and y = T

$$\left(\frac{\partial S}{\partial B}\right)_T = \left(\frac{\partial I}{\partial T}\right)_B \qquad ...(vi)$$

(2) Taking x = B and y = S

$$\left(\frac{\partial T}{\partial B}\right)_S = -\left(\frac{\partial I}{\partial S}\right)_B \qquad ...(vii)$$

From equation (vii)

$$\left(\frac{\partial T}{\partial B}\right)_S = -\left(\frac{\partial I}{\partial S}\right)_B$$

$$= \frac{-\left(\frac{\partial I}{\partial T}\right)_B}{\left(\frac{\partial S}{\partial T}\right)_B}$$

But $T\left(\frac{\partial S}{\partial T}\right)_B = C_B$, the specific heat of the substance at constant field.

$$\therefore \quad \left(\frac{\partial T}{\partial B}\right)_B = -\left(\frac{\partial I}{\partial T}\right)_B \times \frac{T}{C_B}$$

or

$$\left(\frac{\partial T}{\partial B}\right)_S = -\frac{T}{C_B}\left(\frac{\partial I}{\partial T}\right)_B \qquad \text{...(viii)}$$

Equation (viii) gives the change in temperature during adiabatic demagnetisation. This phenomenon, where there is change in temperature due to adiabatic demagnetisation is called *maganetocaloric effect.*

When the field is changed from B_i to B_f, we have

$$T_f - T_t = -\int_{B_i}^{B_f} \frac{T}{C_B}\left(\frac{\partial I}{\partial T}\right)_B dB \qquad \text{...(ix)}$$

But $I = \chi BV$ and $\quad \chi = \frac{C}{T}$ (from Curie law)

$$T_f - T_i = \int_{B_i}^{B_f} \frac{T}{C_B} \frac{d}{dT}(\chi BV)\, dB$$

$$= -\int_{B_i}^{B_f} \frac{TBV}{C_B} \frac{d}{dT}\left(\frac{C}{T}\right) dB$$

$$= -\int_{B_i}^{B_f} \frac{TBVC}{C_B}\left(-\frac{1}{T^2}\right) dB$$

$$= \frac{CV}{C_B T}\int_{B_i}^{B_f} B.dB \qquad \text{...(x)}$$

$$\therefore \quad T_f - T_i = \frac{CV}{C_B T} \cdot \frac{\left(B_f^2 - B_i^2\right)}{2}$$

When the magnetic field is switched off,

$$B_f = 0$$

$$T_f - T_i = -\frac{CVB_i^2}{2C_BT_i} \qquad \text{...(xi)}$$

As $(T_f - T_i)$ is negative, it means there is fall in temperature due to adiabatic demagnetisation. Therefore fall in temperature,

$$\Delta^T = \frac{CVB_i^2}{2\,C_BT_i} \qquad \text{...(xii)}$$

6.11 CONVERSION OF MAGNETIC TEMPERATURE TO KELVIN TEMPERATURE

The thermodynamical behaviour of a paramagnetic crystal placed in a magnetic field is given by the relation

$\delta H = dU$ +s is kept zero and constant, we have

$$\theta = \left(\frac{\partial U}{\partial S}\right)_{B=0}$$

$$\theta = \frac{\left(\frac{\partial U}{\partial T}\right)_{B=0}}{\left(\frac{\partial S}{\partial T}\right)_{B=0}} \qquad \text{...(ii)}$$

Here θ is the temperature expressed on the Kelvin scale and T is the magnetic temperature.

To determine the value of θ, the values of $\left(\frac{\partial U}{\partial T}\right)_{B=0}$ and $\left(\frac{\partial S}{\partial T}\right)_{B=0}$ are to be evaluated. It is done as follows:

The paramagnetic substance is taken at a known initial temperature θ_i. Here θ_i is found using a helium vapour pressure thermometer.

The initial state of the substance, when B = 0, is represented by the point A (Fig. 6.10). The entropy at A = S_1. When the field is increased, keeping temperature constant, the point B_1 is reached. Let S_2 be the entropy of the substance at B_1. The quantity of heat δH produced during magnetisation is directly measured. The change in entropy

$$dS = S_2 - S_1 = \frac{\delta H}{\theta_i}$$

Now, the substance is adiabatically demagentised until the field becomes zero and the point C_1 is reached. During this process there is no change in entropy. The difference in entropy between the points C_1 and A is equal to difference in entropy between the points B_1 and A.

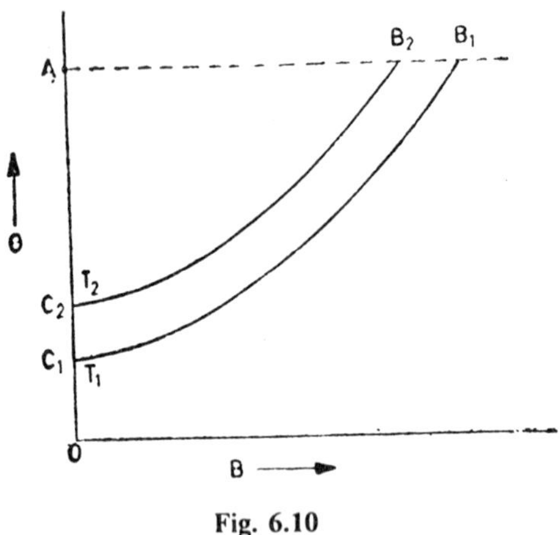

Fig. 6.10

Suppose, the temperature at the point C_1 is θ degrees Kelvin or T degrees magnetic. The values T and $S_2 - S_1$ will depend upon the value of the magnetising field B. The experiment is repeated with different initial fields and a graph is plotted between dS and T (Fig. 6.11).

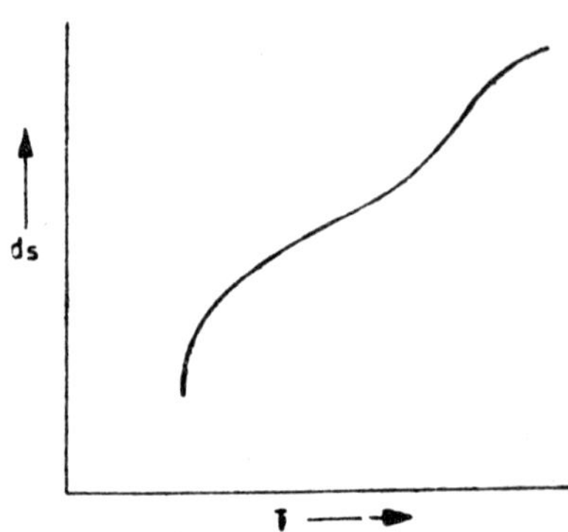

Fig. 6.11

From this graph, the value of $\left(\frac{\delta S}{\partial T}\right)_{B=0}$ is found.

To determine the value of $\left(\frac{\delta U}{\partial T}\right)_{B=0}$, the salt is heated from a temperature T to T' and the heat absorbed is estimated directly. This can be done by introducing γ-rays or radiations from a heated filament.

Substituting the values of $\left(\frac{\delta S}{\partial T}\right)_{B=0}$ and $\left(\frac{\partial U}{\partial T}\right)_{B=0}$ in equation (ii) the value of θ in degree Kelvin can be calculated.

A lower temperature can be reached by using a mixed salt because magnetic interactions between the paramagnetic ions are weakened when the paramagnetic salt is mixed with a non-paramagnetic salt. A mixed crystal can also be used.

Haas and Wiersma (1935) were able to produce temperatures up to 0.003 K using ferric ammonium alum and potassium chrome alum.

6.12 HELIUM VAPOUR PRESSURE THERMOMETER

This thermometer is used to measure temperatures upto 0.7 K. The apparatus consists of a bulb A containing liquid helium. This bulb is connected to the manometer limbs M_1 and M_2 through a connecting tube C. The tube C is surrounded by a copper tube B to ensure uniform temperature of the vapour. R is a reservoir containing mercury (Fig. 6.12).

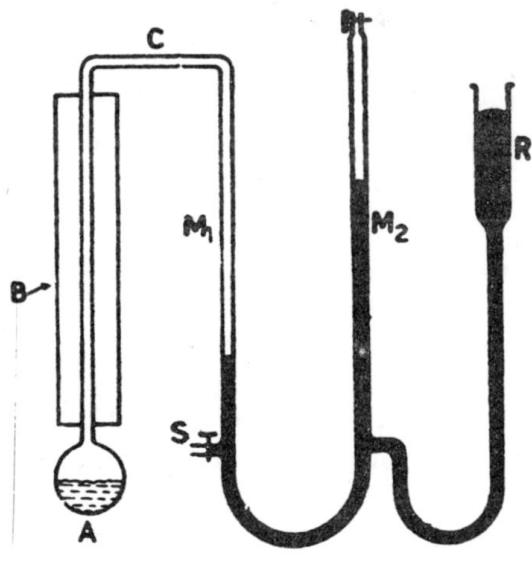

Fig. 6.12

Initially the reservoir R is lowered so that the mercury in the manometers M_1 and M_2 is below the stop-cock S. The tube is connected to an evacuation pump to remove air in the tube C and the bulb A. The stop cock S is closed after evacuation and the bulb A is placed in the bath whose temperature is to be measured. The pressure of saturated helium vapour is measured from the difference in levels of Hg in the limbs M_1 and M_2. With the help of constant tables, giving the vapour pressure of the liquid at various temperatures, the temperature corresponding to any observed vapour pressure is determined. For lower temperatures, the graph between saturated vapour pressure and temperature is extrapolated.

Vapour pressure of Helium (He^4) at different temperature

Temp. K	Pressure mm of Hg
5.00	1460
4.50	980
4.00	615
3.50	353
3.00	181
2.50	77
2.00	23.4
1.50	3.6
1.00	0.12
0.50	1.6×10^{-5}
0.10	3.4×10^{-32}

6.13 SUPER-CONDUCTIVITY

The electrical resistance of metals decreases with decrease in temperature. The decrease in resistance is almost proportional to the decrease in temperature. Before the production of low temperatures, it was thought that the electrical resistance of a conductor becomes zero only at absolute zero. K. Onnes in 1911 performed experiments to determine electrical resistance of pure mercury at the temperature of liquid helium. It was noticed that at 4.2 K the electrical resistance became zero and the metal acquired the property of Super-conductivity. The phenomenon of Super-conductivity opened a new series of problems and scope for new experimentation. It was found that 21 metallic elements and many alloys show the property of Super-conductivity.

The transition temperature (the temperature at which Super-conductivity appears) varies from 8 K for niobium to 0.35 K for Hafnium. Niobium nitride has a transition temperature of 15.5 K. The absence of electrical resistance below the transition temperature is illustrated by the fact that a current of about 1000 amperes passed through a tin wire at about 3 K shows no heating at all. If a current is passed through a ring of lead at a temperature of 7 K (say inductively) the current circulates undiminished through the ring for a number of days.

The recent experiments on Super-conductivity reveal that this property is not confined to a few metals or alloys but may be present in all metals or alloys provided they can be cooled to temperatures nearer absolute zero.

6.14 ELECTROLUX REFRIGERATOR

The schematic diagram of an electrolux refrigerator is shown in Fig. 6.13.

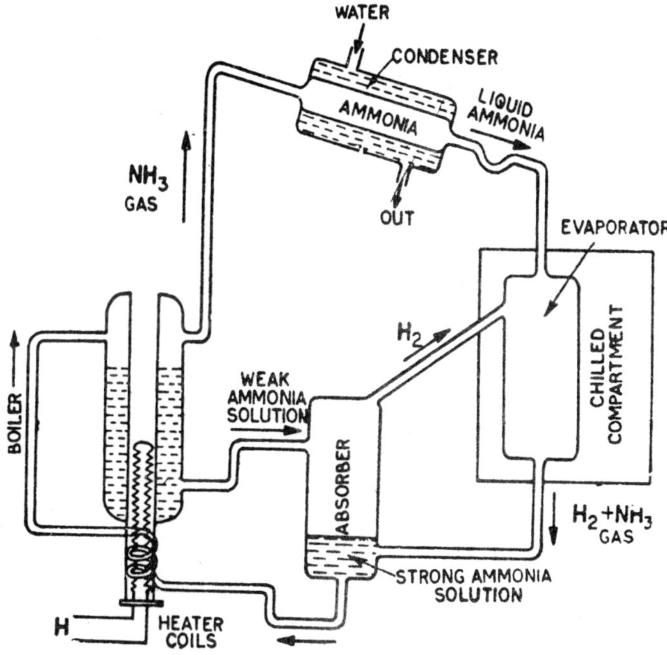

Fig. 6.13

The weak ammonia solution in the boiler is forced into the absorber. The strong ammonia solution goes into the boiler and the gaseous ammonia enters the condenser. It is cooled and is condensed. Liquid ammonia enters the evaporator and is mixed with hydrogen. Hydrogen reduces the partial pressure of ammonia below its saturation point and causes evaporation. This evaporator is surrounded by the chilled compartment. Hydrogen and gaseous ammonia leave the evaporator and enter the absorber. Here they meet the weak ammonia solution. The ammonia gas is dissolved and hydrogen gas rises through the absorber and enters the evaporator. The strong ammonia solution is forced up into the boiler again. The process continues and a sufficiently low temperature is produced in the chilled compartment. The advantage of this apparatus is that no compressor is required and the circulation of the liquid and the gas is automatic.

EXERCISES

1. Give the experimental arrangement of the cascade process for the liquefaction of oxygen.
2. Describe Linde's process for the liquefaction of air.
3. Give the industrial process for the liquefaction of hydrogen. Explain the principle on which it is based.
4. Describe Claude's process for the liquefaction of air.
5. Discuss K. Onnes method for the liquefaction of helium.
6. Discuss the properties of Helium I and Helium II, What do you understand by λ–point? Give the properties of helium II.
7. Write a short account on the production and measurement of low temperatures.
8. Describe the methods for the liquefaction of hydrogen and helium using Joule–Thomson effect.
9. Describe Joule–Thomson effect and give its theory. How has it been utilised in the liquefaction of gases?
10. Describe with necessary theory, the method of adiabatic demagnetisation for producing very low temperatures. How are such temperature measured on absolute scale?
11. Describe with necessary theory the method of producing very low temperatures by adiabatic demagnetisation. Give a method to measure-such low temperatures.

12. Describe fully one method of liquefying a gas. State clearly the principle underlying the method and explain how the temperature of liquefaction may be measured.
13. Explain the phenomenon of adiabatic demagnetisation. How will you employ this phenomenon to produce and measure very low temperatures.
14. Give an account of liquefaction of gases. Explain regenerative cooling method.
15. Distinguish between adiabatic process and Joule–Thomson effect. Show that the Joule–Thomson effect is zero for a perfect gas.
16. Obtain an expression for the fall in temperature due to adiabatic demagnetisation in a paramagnetic gas obeying Curie's Law.
17. Write short notes on:
    (i) Adiabatic demagnetisation.
    (ii) Super-conductivity.
    (iii) Helium vapour pressure thermometer.
    (iv) Helium II.
    (v) Electrolux refrigerator.
    (vi) Production of low temperatures.
    (vii) Liquefaction of gases.
    (viii) Approach to absolute zero.